MEGAFAUNA

MEGAFAUNA

First Victims of the
Human-Caused Extinction

———

Baz Edmeades

HOUNDSTOOTH PRESS

MEGAFAUNA

First Victims of the Human-Caused Extinction

ISBN   978-1-5445-2653-9   *Hardcover*

   978-1-5445-2651-5   *Paperback*

   978-1-5445-2652-2   *Ebook*

Contents

SECTION 3

THE GROWTH OF OUR AWARENESS ABOUT THE ORIGINS OF THE
HUMAN FAMILY AND THE EVOLUTION OF ITS COGNITIVE POWERS

SECTION 1

SHAKING THE FRAMEWORK OF THE GLOBE

Introduction to the deep roots of the human-caused extinction and the unique evolutionary development that set it in motion.

CHAPTER 1

"All the Hugest, and Fiercest, and Strangest Forms Have Recently Disappeared"

EXPLORING THE EASTERN COAST OF ARGENTINA IN THE EARLY 1830s, in the vicinity of what is today the Puerto Belgrano Naval Base, the young Charles Darwin examined the bones of the elephant-sized ground-sloth *Megatherium* and at least nine other extinct species he described as "gigantic quadrupeds."

"It is impossible," he wrote, "to reflect on the changed state of the American continent without the deepest astonishment. Formerly it must have swarmed with great monsters: now we find mere pigmies [sic], compared with the antecedent allied races." Darwin was astounded, too, by how *recently* these "great monsters" had become extinct: their bones were embedded, in some cases, in deposits containing the shells of modern, still-existing mollusk species. It was a state of affairs that cried out for an explanation. "Did man,"

Darwin wondered, "after his first inroad into South America, destroy, as has been suggested, the unwieldy Megatherium and the other Edentata?"

The mystery deepened when his friend and mentor Charles Lyell informed him that North America had also lost a suite of "great monsters" (which included mammoths among many other large mammal, reptile, and bird species). "The mind at first is irresistibly hurried," Darwin wrote, "into the belief of some great catastrophe; but thus to destroy animals...in Southern Patagonia, in Brazil, on the Cordillera of Peru, in North America up to Behring's Straits, we must shake the entire framework of the globe."

Darwin was to learn, in the years that followed his return from the *Beagle* voyage, that this "earth-shaking" wave of large-animal extinctions had engulfed not only the Americas, but also Eurasia (which had seen the relatively recent loss of mammoths, mastodonts, woolly rhinoceroses, giant deer and many other species) and Australia (where a rhino-sized marsupial herbivore, a marsupial "lion" approaching the "real" or placental lion in size, and a great many other big marsupials, reptiles and birds also disappeared near the end of the Pleistocene Epoch).

Alfred Russel Wallace, the co-discoverer with Darwin of evolution by natural selection, was equally puzzled by this recent disappearance of so many of the earth's big land-animals, writing, in 1876, that

> ...we live in a zoologically impoverished world, from which all the hugest, and fiercest, and strangest forms have recently disappeared; and it is, no doubt, a much better world for us now that they have gone. Yet it is a marvelous fact, and one that has hardly been sufficiently dwelt upon, this sudden dying out of so many large Mammalia, not in one place, but over half the land surface of the globe.

At first, Wallace thought that the "sudden dying out" of these big beasts had been caused by "the great and recent physical change known as the 'Glacial Epoch'," but, writing in 1911, his eighty-eighth year, he changed his mind:

> Looking at the whole subject again, with the much larger body of facts at our command, I am convinced that the rapidity of the extinction of so many large Mammalia is actually due to man's agency, acting in co-operation with those natural causes which at the culmination of each geological era has lead to the extinction of the larger, the most specialized, or the most strangely modified forms.

In our time, many students of this recently extinct megafauna have come to agree that the big beasts were victims of a human-inflicted overkill. Others still refuse, however, to accept this conclusion, speculating that the big animals in question may have been killed off by non-human factors such as climate change and disease. I'll try to explain, in the chapters that follow, why I ally myself with people like the paleoecologist John Alroy who believe that the "humans did it" conclusion has long since "been 'proven' as thoroughly as any historical hypothesis can be":

> All of the key evidence was available years ago, and all of it firmly refutes competing, ecologically oriented hypotheses. The event's timing, rapidity, selectivity, and geographic pattern all make good sense according to the anthropogenic model, and no sense at all otherwise. To my eyes, this assessment is so clear-cut that further "tests" are not really necessary.

* * *

Humans did not stop wiping out other species after they completed their settlement of the planet's habitable continents some 12,000 years ago. The number of extinctions caused by our species increased, on the contrary, as we discovered and occupied the earth's previously uninhabited islands. The more accessible of those islands, such as those in the Mediterranean and Caribbean seas, were reached between 10,000 and 5,000 years ago. After they had been occupied by our species, the Mediterranean islands lost a fauna which included several species of dwarf elephants, as well as dwarf mammoth, hippo and deer species, giant dormice, giant eagles, the Majorcan "goat antelope," and a variety of giant tortoise species. After humans reached the Caribbean islands between 6,000 and 3,000 years before the present, those islands also lost giant tortoises, as well as dwarf ground sloths, endemic monkeys, bear-sized rodents, and owl species in a range of sizes characterized by Jared Diamond as "normal, giant, colossal, and titanic."

Humans only completed the task of discovering the planet's more remote islands in the eighteenth century, and the settlement of almost every newly discovered island group appears to have been followed by extinctions. The work of Helen James and Storrs Olson has shown, for instance, that, after Polynesian seafarers reached Hawaii around the beginning of the Common Era, they definitely exterminated thirty-five bird species, and probably exterminated fifty-five in total. After reaching the world's fourth-largest island, Madagascar, at a time calculated to be somewhere between 920 CE and 670 CE, Indonesian seafarers had wiped out, by approximately the end of the fourteenth century, the island's gorilla-sized lemurs and the world's largest bird, *Aepyornis*, among a great many other mammals, birds and reptiles. When Polynesians reached New Zealand in the late thirteenth or fourteenth century, it took them only about a hundred years to exterminate the islands' avian megafauna, which included 500-pound, ten-feet-tall flightless moas and the world's largest eagle, as well as many other large and mostly flightless bird species, while the rats that had accompanied the human settlers multiplied explosively, and killed off an enormous "minifauna" of endemic frogs, flightless songbirds, ground-dwelling bats and large insect species.

As the dodos and their relatives, solitaires, were disappearing during the seventeenth and eighteenth centuries from one of the last-to-be-discovered island groups, the Mascarenes, (along with the endemic pigeons, ibises, rails, owls and some of the giant tortoise species that shared those Western Indian Ocean islands with them), humans were still exterminating big animals on or near the continental land masses. In 1627, the last aurochs, the ancestor of domesticated cattle, was killed in the Jaktoróv Forest in what is today Poland's Warsaw Province, and in 1768—on the eve of both the industrial and the American revolutions—hunters from Kamchatka wiped out Hydrodamalis gigas, an elephant-sized dugong discovered by Vitus Bering's expedition just twenty-seven years earlier around the previously undiscovered Komandorski Islands in the Bering Strait.

In the two-and-a-half centuries that have elapsed since this giant dugong's disappearance, literally hundreds of vertebrate species have followed it into extinction. Nineteenth-century losses of life-forms "charismatic" enough to be noticed by our species included the bloubok antelope, the partially-striped quagga zebra, a Falkland Islands canid, and the great auk. In the twentieth century, extinctions of conspicuous species included the passenger pigeon, the earth's two biggest woodpecker species, two Australian wallaby species, the Caribbean monk seal, and the Tasmanian marsupial "wolf." "Charismatic" or "conspicuous" species wiped out so far in the twenty-first century include the Yangtze dolphin, the six-meter Chinese paddlefish and the 500-pound Yangtze giant turtle. Other large animals are poised to follow. The northern white rhinoceros is represented by only two females as I write this in 2021, while the Javan and Sumatran rhino species have, like many other species, declined to critically endangered levels.

As the population of our species doubled between 1973 and 2021, the process of human-caused extinction intensified. According to the WWF's 2020 *Living Planet Report*, the populations of wild vertebrates (i.e., mammals, birds, fishes, reptiles and amphibians) declined, on average, by a shocking 60 percent between 1970 and 2016. "If there was a 60 percent decline in the human population," wrote Mike Barrett, executive director of science and conservation at WWF, "that would be equivalent to emptying North America, South America, Africa, Europe, China and Oceania. That is the scale of what we have done."

Species other than vertebrates are also suffering steep declines: insect populations in many parts of the earth have been seriously diminished. Invertebrate species represent more than 99 percent of animal diversity, but they don't get their proportionate share of either popular or scientific attention. Non-marine mollusks (clams, mussels, snails, slugs, and limpets) are, for instance, being hit particularly hard by our species' activities, but they don't occupy anything like the space taken up in our consciousness by more "relatable" life-forms like birds.

Humans have also, since 1900, wiped out an average of three plant species per year, "an extinction rate," *Science Alert* notes, that is "at least 500 times faster than is naturally expected…"

Edward O. Wilson, one of the most highly regarded living biologists, warns, moreover, that the already ruinous rate of species-loss is likely to undergo further acceleration in the near future:

> The ongoing extinction rate is calculated in the most conserva-
> tive estimates to be about a hundred times above that prevailing
> before humans appeared on earth, and it is expected to rise to at
> least a thousand times greater or more in the next few decades.
> If this rise continues unabated, the cost to humanity, in wealth,
> environmental security, and quality of life, will be catastrophic.

CHAPTER 2

A Unique and Destabilizing Capability

THE DEVASTATING IMPACT OF OUR SPECIES ON THE BIOSPHERE outlined in the previous chapter has driven some people to the conclusion that humans must be flawed in some way—lacking, perhaps, in prudence or moral restraint. Humans do a great many things that fall well inside their own conceptions of "cruelty" or "greed," but the exceptional intelligence that makes *Homo sapiens* so much more destructive than any other animal, also allows it to experience something that is literally inconceivable for other species: concern for the survival of other life-forms.

In August of 1998, while Jonathan Kathrein was surfing at Stinson Beach near San Francisco, a great white shark seized him by the thigh and dragged him three meters under the water, thrashing him back and forth. When Kathrein somehow managed to get a hold on the edge of one of the shark's gill slits and pull on it, the animal let him go, and he managed to get back on his board and make it to shore with injuries from which he would eventually recover after years of therapy. It would be understandable if this experience left Kathrein with a hostile and even vengeful attitude toward sharks, but it has impelled him, instead, to work for their conservation. It's inconceivable, of course, that a baboon that has survived an attack by a leopard could be motivated to make efforts to protect leopards. It is only "man," in the concluding words of Darwin's *Descent of Man*, "with all his noble qualities, with sympathy which feels for the most debased, with benevolence which extends... to the humblest living creature," that could react in such a biologically anomalous way.

As we'll see in subsequent chapters of this book, our species has been able to save, at least for the time being, a great many mammals, reptiles and birds from the threat of extinction by its own activities. The fact that humans are making these conservation efforts raises an important question: if the enormous and growing extinction spasm we're currently witnessing is not the result of moral shortcomings of one kind or another in our species—if humans are not crueler, say, or greedier than other species—then what other (presumably aberrant) attribute of our species is causing it to exterminate other species at such an unprecedented rate?

* * *

Two pioneers in the field of evolutionary psychology, Irven DeVore and John Tooby, have sought to provide an answer to this question: early in its evolutionary history, they propose, the human family entered what they describe as "the cognitive niche." That entry, DeVore and Tooby argue, would tilt the ecological playing field in our family's favor—give us an advantage over the planet's other life-forms that would turn out, eventually, to be overwhelming.

Before the human family—hominins—entered this "cognitive niche," competition between all the Earth's organisms took the form of relatively slow-paced genetic warfare. A shrub would defend itself, for instance, against being browsed by black rhinoceroses by evolving toxic chemicals in its leaves, to which the rhinoceroses would respond by evolving liver functions that could detoxify those chemicals. Move and countermove in evolutionary struggles of this kind were made over hundreds of thousands, even millions, of years. A worldwide web of such genetic arms races linked each member of the biosphere, directly or indirectly, in a network that Ernst Häckel termed *der Ökologie* in 1866. Although it is pervaded by conflicts of this kind, this network has always had the paradoxical effect of sustaining and (after each of the five mass-extinction spasms and several smaller ones the biosphere has endured so far) restoring the planet's biodiversity.

How did the entry of hominins into DeVore and Tooby's cognitive niche cause them to undo this protective and regenerative function of the ecology? This question will be discussed more fully in Chapter 15 of this book, but for now it will be enough to say that their entry into the DeVore-Tooby niche gave hominins the ability to think up—to invent—useful competitive behaviors and devices, such as making stone blades to cut meat, rather than having to wait the very much longer period of time natural selection would require to assemble meat-cutters for them from chance genetic mutations (as it did, for instance, to equip lions with their meat-slicing or "carnassial" back teeth).

This unique inventive ability gave our family the ability to produce what John Tooby (writing with Leda Cosmides) describes as "... innovations that are too rapid with respect to evolutionary time for their antagonists to evolve defenses by natural selection." "Armed with this advantage," these writers continue,

> hominids have exploded into new habitats, developed an astonishing diversity of subsistence and resource extraction methods, caused the extinction of many prey species in whatever environments they have penetrated...

We didn't ask natural selection to give us this new ability, nor was it in our power to refuse to use it. There could hardly have been a debate among a band of early *Homo sapiens* about whether or not to kindle a fire on a night when they were being threatened by freezing temperatures or encroaching predators. Humans do not, however, have to be spurred on by such life-and-death considerations to reach for the resources made available to them by their inventiveness: they use such resources routinely, instead, for reasons no more pressing than comfort and convenience.

The human family's entry into this cognitive niche did not pose an immediate threat to biodiversity. "How did you go bankrupt?" one of Ernest Hemingway's characters asks another. "Two ways," the other replies. "Gradually, and then suddenly." That's the schedule on which our family's inventive ingenuity, and the consequent power of its technology, have grown. The oldest evidence of hominin-made stone tools found so far, dates to around the middle of the Pliocene, some 3.3 million years ago, and it took nearly a million years for a possible effect of that invention—the extinction of giant tortoises in Africa discussed in Chapter 3—to appear shortly after the beginning of the Pleistocene, approximately 2 million years ago.

Continued increase in the power of our family's technological ingenuity, which was to include, around the middle of the Pleistocene some 1.6 million years ago, control over fire, was followed, approximately 1.4 million years ago, by the relatively slow-moving wave of megafaunal extinctions in Africa discussed in Chapter 4, which engulfed several members of the elephant family, three sabertooth cat species, giant hyenas, giant baboons and a large relative of the giraffe. As a visit to Africa's remaining wilderness areas will show, however, this early extinction episode left a significant number of that continent's big mammal, reptile and bird species untouched. This can be explained by the fact that Africa's fauna co-evolved with the human family, giving many of its members the time they needed to evolve behavioral and other defenses against the relatively slow growth of human inventive power. The South Asian megafauna, which still includes elephants, rhinoceroses, and tigers, was also partially protected from human-caused extinction by early exposure to our family. The biggest mammals, reptiles and birds that inhabited Australia, Northern Eurasia and the Americas were, on the other hand, wiped out relatively completely and relatively abruptly, because humans had, in those cases, entered land masses inhabited by "naive" animals with no evolutionary conditioning to any member of our family, let alone *Homo sapiens*.

* * *

If the extinction-spasm currently eroding the earth's biosphere were in fact the result of greed, cruelty or some other moral shortcoming on our part, then, no matter how difficult it might be, we could at least embark on the task of putting a stop to it by trying to undergo an ethical and behavioral regeneration or reformation. The Canadian environmentalist David Suzuki argues that "aboriginal," or "indigenous" peoples preserved the integrity of their environment by maintaining a "sacred balance" with the natural world. If this belief reflected reality, we could undertake detailed studies of the ethical values and practices of such peoples and set about applying them to our contemporary lives as part of such a regeneration. The real ecological history of the human family presents us, however, with the unfamiliar and disconcerting fact that the anthropogenic extinction-spasm was not set in motion by ethical or spiritual breakdown, but by the morally indifferent evolution of a unique cognitive ability that has given our species an overwhelming advantage over all other members of the biosphere.

Section 2

LOST SERENGETIS

The big mammals, birds and reptiles exterminated by humans and their extinct relatives before the dawn of written history were fascinating creatures in their own right, but the facts of their former existence, and the circumstances of their disappearance, can provide us, additionally, with valuable information about the early history of our own species.

CHAPTER 3

Were Africa's Giant Tortoises the First Victims of Hominin Ingenuity?

"UNLIKE SO MANY PEOPLE WHO GET INFATUATED WITH THEIR own theories," Vance Haynes said of Paul Schultz Martin a few days after he died on September 13, 2010, "he spent his professional career inviting criticism." I got my first chance to see Paul's reaction to that criticism at the American Museum of Natural History's 1997 Extinction Symposium: it was good-natured, courteous and completely free of condescension.

Paul was, in short, the most collegial of men. When I look back over our fifteen-year acquaintance, I can think of only one occasion in which he showed anything like professional envy. That human little moment was an appealing one, though, and I miss him a great deal, now, as I think back on it: I was on the phone with him from Johannesburg, talking about the work of the late Wilhelm Schüle of the Freiburger Institut für Paläowissenschaftliche Studien.

Schüle had been one of the first people to endorse Paul's theory that a member or members of the human family had exterminated several large-animal species in Africa well before the end of the Pleistocene. In doing so, Schüle had adduced a number of facts and arguments of his own. He reasoned, for instance, that Africa's giant tortoises must have been even more vulnerable to growing hominin ingenuity than the large mammals that succumbed to that ingenuity midway through the Pleistocene. It made sense, therefore, (Schüle's argument ran), to expect that Africa's giant tortoises would have disappeared some time *before* that first wave of mammalian extinctions. And that, he pointed out, is in fact what happened: Africa lost its giant tortoise species near the beginning of the Pleistocene.

Partly because some of Schüle's writings are in German, which Paul (his Pennsylvania *Deutsche* antecedents notwithstanding) did not read, and partly because illness had diminished the range of Paul's research by that time, he was hearing about this reasoning for the first time. Before I could get halfway through my account, though, he interrupted with an unexpected "Damn, damn, *damn!*"

Startled, I paused.

"Why," he continued, "didn't *I* think of the tortoises!"

* * *

It's a dry-season morning in the East African Pliocene some 3.75 million years ago, and you're a three-foot nine-inch-tall, sixty-five-pound, female member of Lucy's species, *Australopithecus afarensis*. Under other circumstances you might be enjoying the warmth of this morning sunlight, but, in the last two days, you've had nothing to eat but a small scorpion and some undigested marula nuts you found while sorting through a heap of elephant dung. Last night your milk dried up. At first you were worried that your baby wouldn't stop crying; now you're uneasy about the fact that he hasn't cried for hours.

Suddenly, however, walking a little distance away from the other members of your band, you're hit by a jolt of excitement: a tortoise is plodding along the game trail ahead of you. Lions, with teeth and jaws much bigger and stronger than yours, regularly try and fail to chew open the shells of tortoises of this size, but you may have a technological fix for this problem.

Hominins—members of the human family—may still have been unable to make stone cutting-tools at this time, but they *would* have known—and already have known for a long time—how to use *unmodified* rocks as hammers. Their nearest relatives, chimpanzees, use rock "hammers" to crack nuts. You, in particular, have, after years of clumsy efforts to emulate the experts in your band, learned how to crack open nuts, snails, freshwater crabs, marrow-containing bones, and large eggs with such unmodified stones. You haven't had an opportunity to break open a tortoise's shell yet, but you've watched others doing it, and have, on a few tantalizing occasions, gotten hold of a piece of the delicious, high-fat meat enclosed in that tough protective carapace.

Picking up a tortoise as big as this one is a two-handed job, so you press your baby against your body with one arm—he still has enough energy to clutch onto your chest hair—and lift the reptile up. The tortoise reacts by tucking its head, tail, and legs securely into its shell. Holding the animal, you look around to see if there's anything threatening in the vicinity: memory-flashes of your sister's screams while she was being dragged off by a leopard last winter can still make you wince. Right now, though, there's no sign of danger, so you walk over to a nearby rocky outcrop and carefully place the tortoise upside down in a depression in that rock.

Putting your baby down where you can keep an eye on him, you find a stone the size of a melon, get a two-handed grip on it, lift it over your head, and bring it down as hard as you can on the tortoise's plastron—the armor on the animal's underside. That blow doesn't break the plastron. You have to put down your "hammer," reposition the animal on the rock, and hit it five more times before the story ends well for you (and badly, of course, for the tortoise). By the time the other members of your band come over to see what the noise was about, you've already pulled out the animal's choicest organs, and are happy to let them squabble over the rest of the meat.

Our resourceful tortoise-smasher was lucky to encounter a tortoise that was small by Pliocene standards: at least two tortoise species as big or bigger than the giants presently living on the Galapagos Islands were still living in Africa during her lifetime, and even the combined efforts of the big, five-foot-tall males in her community may not have been able to prevail against the shells of those 500- to 800-pound monsters.

* * *

During the 3.75-million-years-ago lifetime of our tortoise-smasher, the power of hominin inventiveness was, however, on an upward trajectory. In 2010, two ungulate bones dated to 3.4 million years ago with cut-marks apparently made with stone tools, were (as we'll see in several other contexts) found in Ethiopia's Afar Region. Because of the astonishing age of those finds, many paleoanthropologists were reluctant to accept that the marks in question were in fact purposefully made by stone tools, but their criticisms were muted five years later by the discovery, in 3.3-million-year-old deposits, of the first deliberately manufactured stone tools found so far, at a site named Lomekwi 3 on the west side of Lake Turkana in Northern Kenya. These tools were large and clumsily made compared to the "Oldowan" tools that were to appear some three-quarters of a million years later, but they were unquestionably tools.

"[I]t is tempting," writes Terry Harrison, director of the Center for the Study of Human Origins at New York University,

> to infer a direct relationship between the extinction of giant tortoises in Africa with the appearance of early *Homo* and stone

tool using behaviors. During the Miocene and early Pliocene, large size in tortoises would have been an effective strategy to counter predation by carnivores, but by the Late Pliocene, with the appearance of early *Homo* and the use of lithic technologies, natural selection would operate against large size in favor of smaller, wider ranging, faster reproducing and more cryptic species of tortoises.

Hominins were, as we'll see in Chapters 10–12, already cutting the flesh off antelope bones close to a million years before the disappearances of the giant tortoises, and using stone "hammers" to smash those bones to get at their marrow. Combine this with the fact that archaeological evidence, starting at about 1.95 million years BP, shows that tortoise was a popular hominin menu item during the rest of the Pleistocene Epoch, and it's by no means far-fetched to imagine that hominins would, well before the final disappearance of those giant tortoises, already have been using rock hammers to smash open tortoise carapaces.

The late Wilhelm Schüle, the archaeologist affiliated with the previously mentioned Freiburg Institut, investigated the disappearance of the giant tortoise species that had lived on several islands in the Mediterranean until around 8,000 years ago, and concluded that those giants had become extinct soon after humans had first reached the islands in question. The Mediterranean disappearances were, he realized, part of a worldwide tendency for giant tortoise species to become extinct soon after humans or their hominin relatives reached the continents, regions or islands which those reptiles inhabited. He concluded from this pattern, that the disappearances of the giant tortoise species which had lived in Africa during the Miocene and Pliocene Epochs, could have been connected to the rise, on that continent, of the human family.

The extermination of those African tortoises would have taken place, Schüle reasoned, when hominins "adopted a more carnivorous diet during the Upper Miocene or Lower Pliocene." While I agree with Schüle's idea that hominins were probably responsible for the disappearance of the African giants, I don't believe that this disappearance was caused by the rise of, or by an intensification of, meat-eating in our family. As we'll see in Chapter 11, our family probably became carnivorous even before it split off from the chimpanzee line around 7 million years ago. The disappearances of the giant tortoises of Africa resulted, in my view, from an "ontogenetic ambush" constituted by the relatively abrupt development, by a family that had already been carnivorous for millions of years, of the ability to break open the carapaces and plastrons of the largest tortoises with rocks.

* * *

At least two genera of giant tortoises inhabited early Pliocene Africa: *Stigmochelys* and *Centrochelys*. Contemporary estimates suggest that the biggest members of these genera might have exceeded six feet in length, and reached a weight of about 800 pounds.

In India, a member of the genus *Megalochelys* (named in 1837 by Falconer and Cautley from Plio-Pleistocene deposits in the Siwalik Hills) measured up to ten feet from the front of its carapace to the back, and reached some 2,000 pounds in weight. This monster survived, it seems, somewhat longer than Africa's giants; it was still around for some time into the Pleistocene. Other giant tortoise species—possibly also members of *Megalochelys*, although their taxonomy has not yet been sorted out satisfactorily—inhabited Sumatra, Java and other East Indian islands in the early Pleistocene. (As we'll see presently, hominins entered the southern parts of Asia some 2 million years ago.) Giant tortoises living on Flores Island, which wasn't joined to the Asian mainland by the falling sea-levels of the Pleistocene glaciations like nearby Java and Bali were, survived until about 800,000 years ago, when members of the human family managed to settle Flores by crossing the water-barrier—ten to fifteen miles wide during the height of the glaciation— which separated it from the mainland.

Evolution enlarged the front and back shell openings of the extinct "saddle-back" species from Rodrigues Island in the Mascarenes, and of a few varieties of the Galapagos species. This increased the reach of their heads for browsing purposes and gave their legs more mobility. Species with modifications of this kind could have been vulnerable to large continental predators like lions, bears and hyenas. Giant tortoises such as those living on the African, Eurasian and American continents, which retained "normal-sized" shell-openings (i.e., relatively small ones), would, on the other hand, have been secure from such four-legged predators, despite the fact that they would not have had the power to flee from, or harm them.

They would, however, only have had one line of defense against animals wanting to gain access to their meat: the fact that they could withdraw into an impregnable carapace. "I was always amused," Darwin wrote about the Galapagos giants,

> when overtaking one of these great monsters as it was quietly pacing along, to see how suddenly, the instant I passed, it would draw in its head and legs, and uttering a deep hiss would fall to the ground with a heavy sound, as if struck dead.

Exclusive reliance on this defense would, sooner or later, have made these animals vulnerable to a predator who was starting to explore the world of inventive innovation. By exposing a rich store of meat to view, and protecting it only by enclosing it in a hard structure, nature was setting the Pliocene predators of Africa the same kind of intelligence test as primatologists set when they hang bananas from ceilings to see if chimpanzees can reach them by manipulating boxes and poles. As long as no member of Africa's Pliocene predator-guild could pass that test, that continent was as viable an environment for giant tortoises as any oceanic island.

Baboons in South Africa's De Hoop Nature Reserve have been seen to eat tortoises belonging to the small *Chersina angulata* species, but they use their hands and teeth to open the relatively fragile shells of that species' juveniles. Chimpanzees have recently been observed breaking open tortoises' plastrons by picking the animals up and pounding them, belly first, against hard tree trunks and branches in Gabon's Loango National Park. (Some, but not all, West African chimpanzees use hand-held rock hammers to break open *nuts*, but no chimpanzee has, as yet, been seen using that technique on tortoises.) It seems safe to say, however, that breaking open a giant tortoise's armor—a task that would have required the purposeful, two-handed manipulation of relatively large rocks for an extended period of time—is far beyond the capabilities of any primate outside the hominin family.

It's highly likely, on the other hand, that this task was already *within* the power of beings who had developed the ability to use hammer-stones to knock sharp cutting flakes off other rocks, and use those flakes to butcher horse-sized mammalian carcasses. As my opening vignette suggests, smaller tortoises probably became vulnerable to our family before any of its members learned to make stone cutting tools. As we'll see in later chapters, hominin tool-use—including the use of unmodified rocks—must have come into existence long before our family was manufacturing stone implements.

We also know, from archaeological finds in Africa, Asia, Europe and the Americas, that tortoises were a valued source of meat for hominins throughout the Pleistocene. The relatively fatty flesh of those animals would, pound for pound, have been richer in calories than that of most other available meat. Judging by accounts written in our era, it would probably have tasted good too. William Dampier, an English pirate-naturalist based on the Galapagos during the seventeenth century, wrote that the meat of the tortoises of those islands was "...so sweet, that no pullet eats more pleasantly."

* * *

Hominins may have started their tortoise-eating career by opening the shells of very young tortoises with their teeth, the way present-day baboons do. From there, they could have started breaking open the carapaces of somewhat larger individuals with "crude" percussive technology—that is, by smashing the tortoise itself against a rock, the way some Gabon chimps smash tortoises against tree trunks. It may have taken some time to move from that method to the point where they began to use rock "hammers" to break into the armor of bigger tortoises.

Although ontogenetic innovations can literally arise overnight, the mental machinery that makes such discoveries and inventions possible, must (as we'll see in Chapters 12 and 15) have been assembled, by natural selection, over thousands of generations. The hominin ability to break open the carapaces of giant tortoises and terrapins with tools would, therefore, have manifested itself in a *relatively* gradual way. One or more of Africa's giants might, therefore, have had time to evolve a smaller body size in response to hominin predation. Size-reduction can—if a species has time to undergo it—be a very effective countermeasure against the threat of human-caused extinction: small tortoise species are, as we saw previously, much less vulnerable to such extermination because they can reproduce more rapidly, exist in larger numbers, and hide more effectively.

I think it's possible, for these reasons, that Africa's biggest present-day tortoise, *Centrochelys sulcata*, the spur-thighed tortoise of the Sahel (which can weigh up to 200 pounds in exceptional cases) and/or *Stigmochelys pardalis*, the leopard tortoise of South-Eastern Africa (which normally reaches 40 pounds, but has been known to get to 90) could possibly be dwarfed descendants of the giant *Centrochelys* and *Stigmochelys* species of the Pliocene. Size-reduction among Africa's tortoises seems to have continued during the Pleistocene, even though the giants had already disappeared by the beginning of that epoch. Richard Klein and Kathryn Cruz-Uribe report that "[t]he tortoises tend to be much larger in the MSA [Middle Stone Age] layers than in the LSA [Later Stone Age] ones..."

Although some or all of the African giant tortoises and terrapins may have escaped extermination at the hands of our family by dwarfing, there's no doubt that the fate of the overwhelming majority of giant tortoise species exposed to our species has been outright extinction. The giants that still lived on dozens of islands after their continental counterparts had disappeared are represented today by only eleven subspecies of *Chelonoidis nigra*, found in the Galapagos, and by two or three closely related members of the genus *Dipsochelys* which survive on islands forming part of the Republic of Seychelles.

By the beginning of the Common Era, a rich radiation of giant tortoise species still extended across Madagascar, the Comoros, the main Seychelles islands, the Aldabra group, the Glorieuses, the Amirantes and the Mascarenes. The surviving Indian Ocean species are geographically and genetically widely separated from their Galapagos counterparts, but they reach approximately the same size: their carapaces are about four feet long measured over the curve, and they commonly reach 500 pounds in weight. Esmeralda, who lives on Bird Island near the main Seychelles (and happens, incidentally, to be male), is claimed to be the world's largest tortoise. He's thought to be around 180 years old and weighs about 800 pounds.

The reason why tortoises inhabited so many islands in the planet's seas and oceans is that they can remain afloat and alive for extraordinary periods of time without access to food or fresh water. Their longevity means, moreover, that, after making landfall on an island, they can wait around for a century or more for a mate to be washed up by the same current that brought them to their new habitat (although many colonizations probably started by the arrival of a single female carrying fertilized eggs). Some of the tortoise species that colonized islands in this way may not have been giant-sized when they first arrived—enlargement of their species would have taken place, instead, in a relatively rapid way, after their arrival. On relatively small islands, factors such as reduced predation lead to a relatively rapid size-increase in small species, and a decrease in the size of big ones like elephant or deer.

* * *

Could climate change have killed off Africa's giant tortoises? There's no question that many tortoise species, large and small, must have been exterminated by this agency. The giant *Meiolania* tortoises and the side-necked turtle species that inhabited Antarctica early in the Cenozoic or "age of mammals," and the soft-shelled turtles that lived on Greenland and on other islands in the vicinity of the North Pole at that time, were clearly victims of the present Ice Age. Many tortoise and terrapin species would not have had the option of retreating away from the poles to escape the temperature drop that the present Ice Age was bringing to the higher latitudes. The fact that giant *Cheirogaster* tortoises disappear from Northern Europe by about 5 million years ago, but survive in the southern regions of that continent until at least 2 million years ago, must also be the result of high-latitude cooling. It's hard to understand, however, how climate change could have exterminated *Cheirogaster* in the *southern* parts of Europe. Smaller tortoises are, after all, still widespread and relatively abundant there, and there's no evidence that large tortoise species would have been more vulnerable to cold than these surviving species are. Even though giant tortoises drop off the paleontological radar screen in Southern Europe around 2 million years ago, it's not inconceivable, therefore, that they could have survived there until hominins made their first appearance in that region a million years ago or more.

It's very unlikely that climate change was responsible for the disappearance of giant tortoises whose ranges included the relatively warm regions of Africa, Eurasia and the Americas. Those regions became only marginally cooler during the cold phases of the glacial cycles. The biomes inhabited by tortoises or terrapins—whether jungle, savanna, or desert—grew, shrank, and/or shifted in response to those cycles, but none of them ceased to exist. The Pliocene disappearance of the African giants was not, moreover, accompanied by the extinction of large tortoise species on any of the other warm-climate landmasses: the South Asian species only disappeared, as we've seen, in the early Pleistocene, and the Americas did not lose their giant tortoises until the end of that epoch between 10,000 and 15,000 years ago when most of the New World's megafauna disappeared. The two giant tortoise species that inhabited Madagascar, whose climate is closely tied to that of Africa, disappeared in historical times, following the first human settlement of that island. Nor was the disappearance of the African giants near the beginning of the Pleistocene part of any wider extinction-spasm in Africa itself, whether of tortoises or any other organism.

* * *

The only direct evidence we have of humans exterminating giant tortoises is found in the historical accounts of tens of thousands of tortoises being killed for food by sailors in the last few hundred years in the Seychelles, the Mascarenes and the Galapagos. Taking advantage of these animals' ability to live without food or water for extended periods of time, sailors would bring large numbers of them on board alive, often stacking them upside-down in the ships' holds to kill them later as their meat was required.

The evidence implicating our family in the giant-tortoise extinctions, which took place before this time, is circumstantial, but that doesn't necessarily mean that it's insufficient. In a classic 1894 case that helped to build the Anglo-American law of evidence, *Makin v. Attorney General of New South Wales* (an appeal from an Australian court to the British Privy Council), a man and his wife were accused of murdering a child whose mother had paid them to take him into their care. The child's corpse had been dug up in the yard of a house they'd been renting. The Makins' defense was that the child had died naturally, and that "if they were guilty of anything, it was merely of having improperly buried the child." That plea might well have succeeded if the court hadn't admitted evidence that no less than eleven corpses of other children had been dug up in the yards of various houses occupied by the accused.

Without being told about the other eleven corpses, a jury might have decided that there was, say, a one-in-ten chance that the child in question could have died naturally. That may have been enough to constitute, in the minds of those jurors, the "reasonable doubt" that the criminal law requires for an acquittal. Add one more corpse, however, and the one-in-ten chance I've assigned to the possibility of a single child dying naturally becomes one in a hundred. Dig up a further ten, and it reaches one in 1,000,000,000,000. The Privy Council had no trouble affirming, at any rate, that evidence of the other eleven corpses was admissible. The law is normally cautious about admitting evidence of previous wrongdoing with which the accused wasn't charged ("similar fact evidence"), because of its potential to prejudice the jury against the accused unfairly, but such evidence can be admitted, the Privy Council decided in the *Makin* case, if its probative value outweighs that potential for prejudice.

The frequent recurrence of the "hominins arrive, giant tortoises disappear" sequence that impressed Wilhelm Schüle, establishes a pattern whose probative power is, in my view, as irresistible as the "Makins take in children, children die" sequence. The former sequence manifests itself for the first time when hominins arrive in South Asia around 2 million years ago. It is repeated when they reach Flores Island 800,000 years ago, and reappears when *Homo sapiens* reaches Australia some 50,000 years ago. It is seen again when humans enter and settle the New World around 15,000 years ago; when they reach the previously inaccessible islands of the Mediterranean Sea between 7,000 and 8,000 years ago; when they reach the Caribbean islands some 6,000 years ago; when they make it to the Canaries off the northwest coast of Africa at about the same time; when they sail to New Caledonia Island, some 800 miles east of Australia about 3,500 years ago; when they reach Vanuatu at about 3,250 years ago; when they land on Fiji 2,500 years ago; when they settle Madagascar at about the same time; and, finally, when Arab, Persian, Portuguese, Dutch, French and English ships start making landings on the last undiscovered Indian Ocean islands in the centuries preceding the seventeenth.

These disappearances establish beyond any reasonable doubt that the hominin family was responsible for the extermination of the vast majority, if not all, of the giant tortoise species that inhabited territories newly settled by its members. They show clearly, therefore, that giant tortoises, with their single line of defense against predation, are enormously vulnerable to extermination at the hands of our family. That oft-demonstrated vulnerability, coupled with direct archaeological proof that hominins have found tortoise meat a tasty, easy-to-utilize food package for nearly 2 million years, constitute a strong indication that the giant tortoise species that disappeared from Africa near the Pliocene-Pleistocene border 2.5 million years ago, could have been exterminated and/or dwarfed by hominins, as one of the consequences of the early steps taken by our family into the terrain (touched on in Chapter 2 and discussed more fully in Chapter 15) of cognitive innovation.

CHAPTER 4

The Mid-Pleistocene Depletion of Africa's Megafauna

IN THEIR *ELEPHANTS OF SAVUTI,* ONE OF THE EXCELLENT WILDLIFE videos that Dereck and Beverly Joubert have filmed for *National Geographic,* a half-grown elephant wanders away from its herd and gets surrounded by lions. The Jouberts record the realities of Africa's wilderness ecologies in an unsentimental way, and I expected to see the elephant being killed and eaten in short order. The lions were literally all over the young animal, but were, it turned out, unable to penetrate its skin. It was as though a bunch of kids was trying to bite into a watermelon that hadn't been sliced open—their teeth simply couldn't get a purchase. Even though the cats tried vulnerable places like the elephant's anus, they just couldn't get through that fortress-like animal's outer "wall." In the end, they gave up and the elephant was eventually able to rejoin its herd.

Lions can and do kill elephants on occasion, but it would obviously be much easier for them to do so if they had teeth that were specially evolved to penetrate the skins of animals of that size. As we've seen in several other contexts, early-Pleistocene Africa had no fewer than three big cats whose teeth were specially enlarged and length-ened to penetrate thick skins. These cats—members of the so-called machairodont sub-family—were the saber-tooth *Megantereon*, a "scimitar-tooth" called *Homotherium*, and a "dirk-tooth" known as *Dinofelis*.

Megantereon was about the size of a present-day jaguar but more powerfully built than this still-existing cat. Its front legs, shoulders and chest were, in particular, massively developed. That, and the fact that it could open its jaws much wider than any modern cat can, suggests that it may have used its front limbs, together with its powerful neck muscles, to push or pull its teeth into vulnerable parts of its prey like large blood vessels or the trachea. *Megantere-on*'s stocky build tells us that it was probably an ambush hunter that only pursued its prey for short distances. It had a short, bobcat-like tail. Its six-inch upper canines were laterally flattened so that they could slice through muscle and skin.

Bigger than *Megantereon*, the "scimitar-tooth" *Homotherium* was about the size of a lion, but more slenderly built and possessed of a long, powerful neck like that of a spotted hyena. Also like the hyenas, it had long forelimbs, relatively short back ones, and a short tail. *Homotherium* seems to have specialized in running pursuit. Like the cheetah, it had sacrificed the ability to retract its claws in favor of getting better traction. Like the cheetah, too, it had large nasal open-ings—presumably to allow quicker oxygen intake—and a large and complex visual cortex thought to be associated with daytime hunting.

Homotherium's slender upper canines were shorter than those of *Megantereon* and more curved. The rear edges of those canines were crenulated like the cutting edge of a steak knife. Lack of wear on the upper canines on both *Homotherium* and *Megantereon* suggests that those teeth were used primarily for the business of killing. The tearing and cutting involved in feeding were done, in both species, by very robust incisors, situated much farther forward than those of pantherine cats like lions and tigers, along with the forward-shifted lower canines, and well-developed meat-cutting or carnassial teeth in the back of the jaw.

In a 1970 article in the *Zeitschrift für Säugetierkunde*, Vratislav Mazák, a Czech zoologist, drew attention to the fact that a 30,000- to 35,000-year-old, ten-inch-long stone carving found at Isturitz in the French Pyrenees which had been thought, up to that time, to represent a lion could well be a representation of *Homotherium*.

Reproduced from "On a Supposed Prehistoric Representation of the Pleistocene Scimitar Cat, *Homotherium fabrini*, 1890 (Mammalia; Machairodontidae)" by Vratislav Mazák, in 1970 Zeitschrift für Säugetierkunde, p. 359, at p. 360.

As this drawing shows, the now-lost carving depicted a lightly spotted cat with a short tail and a "swollen" lower jaw (*Homotherium*'s upper canines didn't protrude from its mouth; they were housed, instead, in tough, protective tissue associated with the bone flanges that deepened its bottom jaw). Mazák's suggestion was largely ignored until the carbon dating, in 2002, of a *Homotherium latidens* lower jaw dredged from the North Sea, to about 28,000 years ago. That confirmed the fact that this species was still living in Europe after the first members of *Homo sapiens* entered that continent.

Homotherium was a long-lived, successful genus. After it disappeared from Africa in the 1.4-million-year-ago extinction-spasm we'll discuss in this chapter, it survived in the northern parts of Eurasia, and in North America, until near to the end of the Pleistocene. The Friesenhahn Cave in Texas, in which the remains of 20 adults and 13 juvenile members of *Homotherium serum* were found, also contained the remains of between 300 and 400 juvenile mammoths. The majority of the individuals represented by these remains were around two years old—an age at which modern elephants begin to assert a degree of independence from their mothers. This may give us an idea about the feeding strategy that *Homotherium* could have pursued.

Africa's third machairodont, the "dirk-tooth" or "false saber-tooth" *Dinofelis*, was, like *Megantereon*, a jaguar-sized animal. The combination of powerful forequarters and relatively gracile rear limbs, suggest that it was, also like *Megantereon*, an ambush killer. *Dinofelis*'s upper canines weren't laterally flattened like those of *Megantereon* and *Homotherium*. They were conical, like those of lions and leopards, and midway between those of the lion and *Homotherium* in length. Analysis of its tooth enamel tells us that this cat lived on animals who ate C4 grasses (i.e., grasses that thrive in warm, sunny conditions). *Dinofelis* was, therefore, probably a savanna dweller.

Savannas became permanent features of the African landscape between 5 and 10 million years ago. Africa's main antelope families arose in response to this development. Around 2.5 million years ago the earth plunged to a new low of cold and aridity. This doesn't mean Africa became a frigid place—average temperatures would only have declined by 1 or 2 percent. It does mean, however, that precipitation declined further, expanding savannas at the expense of jungle and forest. Because grasslands are, counterintuitively perhaps, considerably more productive than jungles, antelopes reached new heights of abundance and diversity in Africa. This development represented a big increase in species that relied on swiftness rather than large size and thick skins to avoid being eaten by a predator. "Pantherine" cats like lions and leopards, and felids like cheetahs, all of which had already been in existence for some time, rose to prominence in response to it. Those predators were fast and/or agile enough to catch the swift new grass-eaters, and their teeth didn't have to be particularly big to penetrate the relatively thin skins of these newly important kinds of prey.

Machairodonts must, of course, have competed with pantherines to a degree, but neither group was able to push the other out of existence. "What is now abundantly clear," the fossil-cat expert Alan Turner tells us, is that

> the machairodont cats co-existed in Africa with the modern feline cats for a very long period of time, considerably reducing the force of the argument that their demise may have resulted from competition with the modern species.

The machairodonts would, as a group, have lived off larger, thicker-skinned animals than the pantherines did, and each of the three machairodont species must have occupied more or less separate "sub-niches" within that "larger prey" category. But how many large-prey sub-niches, you might ask, could Africa's biggest herbivores offer? That continent only has one elephant species, after all, along with two rhino and two hippo species.

The answer is that early-Pleistocene Africa was much richer in megaherbivore species than the Africa of today is—so much richer, in fact, that a person familiar with Africa's present-day fauna would be astounded if she could take a time-trip to the Africa of 1.5 million years ago. Her *Collins Guide to African Wildlife* would have to be quite a bit thicker than it is now. It would, in particular, have to have an entire chapter devoted to the different kinds of elephants she'd be encountering. That chapter would tell her that the diverse collection of elephant species that inhabited Africa at that time were divided into two suborders: the Deinotherioidae and the Elephantoidae.

The Deinotherioidae contained only one family, and that family was, as far as we can tell, represented only by the genus *Deinotherium* which contained, in turn, only one species that we know about: *D. giganteus,* an animal somewhat larger than a present-day African elephant, equipped with downward-curving tusks growing out of its bottom jaw. The other proboscidean suborder, the Elephantoidae, consisted of three families: the Gompotheriidae, the Mastodontidae and the Elephantidae. We're not sure if the first two of those families contained more than one genus, but we know that the third—the Elephantidae—contained three: *Loxodonta,* the genus of the surviving African species; *Elephas,* that of the surviving Asian species; and *Mammuthus,* that of the extinct and presumably hairless African mammoths.

Early-Pleistocene Africa was, therefore, inhabited by at least six genera of elephant-like animals. How many species did those six genera contain? Nancy Todd of the Department of Biology of Manhattanville College thinks that the genus Elephas might have been represented, at that time, by as many as five co-existing species. *Loxodonta* may also have been represented by more than one species at a time during the Pleistocene. It is, indeed, still variable enough to create doubt about whether it should be viewed as a single species. On the conservative assumption that each of early-Pleistocene Africa's six proboscidean genera was represented, on average, by 1.5 species, we can conclude that the continent may, at that time, have been inhabited by some nine species of elephant. There could well have been more.

Africa was also, at this time, inhabited by at least four kinds of hippopotamus. The genus *Hippopotamus* had two species, the still-living *H. amphibius,* and a variant with more extreme aquatic adaptations, such as periscope-like eyes and snorkel-like nostrils, called *H. gorgops*. The pig-like genus *Hexaprotodon* had at least two: the still-living *H. liberiensis*, the pygmy hippopotamus, and one or more larger species that are thought to have preferred, like the pygmy member of this genus still does, to spend most of their time on land, in forest cover.

Africa had two rhino species in the early Pleistocene, as it does today, but it was also inhabited by a large relative of the rhino-tapir-horse order (Perissodactyla) known as *Ancylotherium*. This perissodactyl had a ground-sloth-like body, with short, powerful hindquarters and long, muscular "arms" which it seems to have used in conjunction with claw-like hoofs to pull branches within reach of its mouth. The elephants, hippos, and big perissodactyls we've just talked about would have been the most visible part of the continent's fauna, making up perhaps 90 percent of its mammalian biomass.

Pleistocene Africa's medium-sized animals would have been overshadowed by these giants, but the members of that group were also bigger and more diverse than their present-day counterparts. *Metridochoerus* was a giant version of the present-day warthog, while *Colpocheorus* was a giant bush-pig. Africa was, during the Pleistocene, also inhabited by the giant member of the wildebeest-hartebeest family, *Megalotragus*; a giant relative of the still-living roan and sable antelopes named *Hippotragus gigas*; a large zebra known to science as *Equus capensis*; a large, big-antlered member of the giraffe-okapi family, *Sivatherium*; and two huge baboon species—one as big as a gorilla and the other about the size of a modern human.

The vast majority of African mammals that disappeared during the Pleistocene were distinguished by their large size. The ground-dwelling *Paracolobus*, which vanished near the beginning of that epoch, was Africa's largest monkey. Only a few of the vanishing mammals were the same size as their present-day relatives: *Makapania*, which was allied with the still-existing musk ox and takin was one, as was Africa's three-toed horse *Stylohipparion.* Both the "gracile" and "robust" australopithecine members of the human family, which disappeared during that time, were smaller than *Homo erectus.* Only one small antelope that we know of disappeared from Africa during the Pleistocene: a steenbok-like animal assigned, possibly in error, to the springbok genus as *Antidorcas bondi.*

* * *

When did the various species of this extinct African fauna make their last appearances? Giant tortoises were, as we saw in the previous chapter, already dwarfed and/or exterminated near the Pliocene-Pleistocene border. The last of the gracile australopithecine fossils date from about 1.7 million years ago—that is, at roughly the same time as *Homo erectus* made its first appearance.

About 1.4 million years ago a much larger wave of extinctions hit Africa. All of the continent's elephant genera except *Elephas* and *Loxodonta* were swept away by it. (*Loxodonta* survives, of course, in present-day Africa; *Elephas* only disappeared from that continent about 125,000 years ago and survives in South Asia.) One or more of Africa's *Hexaprotodon* or "forest hippo" species, as well as its big ground-sloth analog *Ancylotherium,* were also annihilated in this early-Pleistocene extinction-spasm. This huge drop in the diversity of Africa's megaherbivores was accompanied by the disappearance of the lion-sized hyena *Pachycrocuta,* the cheetah-like "running hyena" *Chasmaporthetes,* all three of that continent's machairodont or "sabertooth" cats, as well as that of a large, poorly identified dog that could have been related to the present-day African wild dog.

This 1.4-million-year-ago hemorrhage was the most destructive extinction episode that Africa was to suffer during the Pleistocene, but the diversity of that continent's big-animal species would, during the rest of that epoch, continue to bleed away at a reduced level. Africa's giant pigs, its "super-aquatic" hippo *H. gorgops,* and its hipparions would disappear around 900,000 years ago. Near this point, the last of the australopithecines, the so-called "robusts," also disappeared from Africa—and from the planet—leaving the genus *Homo* as the only representative of the formerly diverse human family.

In the vicinity of 500,000 years ago, Africa's giant *Hippotragus* antelope, its giant baboon species, and the *Sivatherium* giraffe/okapi slipped into extinction. As we've already seen, *Elephas*, the animal now known as the Asian elephant, became extinct in Africa around the time of the second-last interglacial, the Eemian, which occurred around 125,000 years ago.

At or near the end of the Pleistocene, around 12,000 years ago, Africa's megafauna experienced a final extinction episode that coincided with the catastrophic disappearance of the bulk of the megafauna of both Eurasia and the New World. This final spasm swept away the giant wildebeest *Megalotragus*, the large zebra *Equus capensis*, the African musk-ox *Makapania*, and the small steenbok-like antelope which I referred to as *"Antidorcas" bondi*. The long-horned buffalo *Pelorovis antiquus* also disappeared from Southern Africa at this time, but it seems to have survived on the savannas of the interglacial Sahara until about 5,000 years ago.

The only region other than Africa to experience significant losses of big-animal species in the earlier part of the Pleistocene was South Asia. We'll talk about these early South Asian losses in the next chapter.

* * *

The late Paul S. Martin of the University of Arizona was the first person to present a logical argument for the proposition that hominins caused the spate of big-animal extinctions that took place in Africa around the middle of the Pleistocene Epoch.

In an article published in *Nature* in 1966 titled "Africa and Pleistocene Overkill," Martin contrasted that early diminution of Africa's
megafauna well over a million years ago, with the sudden disappearance of the American "super-Serengeti" around 11,000 years
ago, and the extinction event that destroyed Madagascar's megafauna some 1,500 to 500 years ago. The fact that those three extinction episodes were widely separated in time, excluded, Martin
pointed out, the possibility that they could have been caused by a
single, worldwide episode of climate change. Even the seemingly
significant fact that the American extinctions had taken place at
the same time as the abrupt termination of the last glaciation did
not support a causal link between those two occurrences: climatic
events as severe as the latest glacial termination had, Martin
argued, been taking place throughout the Pleistocene, and none of
them had diminished the big-animal diversity of North America to
any noticeable degree.

But if climatic turbulence associated with glacial cycles had visited
North America on numerous occasions during the Pleistocene,
humans had not. Members of our family arrived in North America
for the first time near the end of the Pleistocene, and the collapse
of the American megafauna took place relatively soon after that
arrival. The same sequence of events had, Martin pointed out,
taken place in Madagascar: relatively soon after humans reached
that island for the first time in the vicinity of 2,000 to 1,500 years
ago, Madagascar lost, among other species, at least six kinds of
"elephant bird," the largest of which was about three times heavier
than an ostrich, seven genera of lemurs of which the largest species
weighed as much or more than a male gorilla, two species of giant
tortoise, and two kinds of hippo.

Many people had, before the appearance of Martin's 1967 article,
used the example of Africa to refute the idea that human hunters
could have caused extinctions such as those that occurred on Madagascar and in the New World. Martin explained this reasoning as
follows:

> If Early Man was responsible for the destruction of the New
> World fauna, the argument goes, why did the evolving hominids
> leave the African fauna intact during their evolutionary devel-
> opment over more than a million years?

Martin's article then went on to inform his readers of the state of
affairs we've just been discussing: that the present-day African
fauna was in no sense "left intact." In the 1960s (early days, as far
as geochronology is concerned) paleontologists were still vague
about the timing of events like the Pleistocene extinctions of many
of Africa's biggest animals, and the advent of Acheulean technology.
Martin's research had, however, given him an accurate idea of the
kinds of animals that Africa had lost during the Pleistocene, and
informed him of the fact that the vast majority of those losses had
taken place well before the end of that epoch.

No extraordinary losses of big-animal species were taking place
elsewhere on the planet at that time (except, as would later become
apparent, in South Asia, the only other landmass then inhabited
by hominins). Madagascar in particular, whose climate is closely
tied to that of Africa, did not experience any discernible increase in
extinctions during the "hemorrhage" that bled away most of Africa's
elephant species and many of its other species of megafauna. For
Martin, the only remaining explanation for that hemorrhage was
a "cultural" one—that is, one that involved hominins and, presumably, the unprecedented degree of power that our family was developing. He pointed out that a new technology, the Acheulean, had
emerged in Africa near the time when Africa was losing many of
its big animals. (We've learned, since Martin made this argument,
that the "Acheulean Revolution," I refer to in Chapter 14, included
discovery of fire-use.)

The timing of the latest two of the three extinction episodes Martin
was contrasting in his 1966 article supported the notion that they
had been caused by the spread of our family across the planet: the
American extinction-spasm happened soon after *sapiens* reached

North America via the Bering Land Bridge, while the collapse of the Madagascan megafauna took place thousands of years later, after our species became capable of reaching the planet's last uncontacted islands by undertaking long sea voyages.

The differing degrees of intensity of those three extinction-spasms also supported the idea that they were caused by our family: extinctions were relatively moderate in Africa, heavy in America, and extremely heavy in Madagascar. This was consistent with the likelihood that the African fauna, which had evolved in company with hominins, would have been more resistant to the advent of hominin ingenuity than the American fauna was, while Madagascar's big animals and birds—which had no exposure at all to any kind of large, sophisticated predators before the arrival of humans—would have been even more vulnerable to humans than those of the Americas.

* * *

After returning to the topic in a chapter titled "Prehistoric Overkill: The Global Model" in the 1984 *Quaternary Extinctions* that he co-edited with Richard Klein, Paul didn't write anything further about the African Pleistocene. The body of work that has, over the last fifty years, established him as a pioneer in the field of Pleistocene extinctions, is focused, for the most part, on the end-Pleistocene disappearance of the North American megafauna.

Since Paul first mooted the question in 1966, relatively little has been written about the possibility that humans caused the early-Pleistocene African extinctions. That's not to say it wasn't on anyone's mind. Lars Werdelin, a curator at the Swedish Museum of Natural History in Stockholm, supports the idea that hominins caused the extinctions of the large carnivores that disappeared from Africa in the earlier Pleistocene. "The way I see it," he said at a 2013 symposium on human evolution and climate change at Columbia University's Lamont-Doherty Earth Observatory in Palisades, New York, "this is one of the first ways in which we manipulated our environment on a large scale."

Todd Surovell, Nichole Waguespack and P. Jeffrey Brantingham provide support for the idea that humans exterminated proboscideans in mid-Pleistocene Africa. As far as I know, however, the only person to fully endorse Paul's idea that hominins were responsible for most of the early-Pleistocene disappearances of African megafauna, was Wilhelm Schüle, the archaeologist whose thoughts on the disappearances of Africa's giant tortoises we discussed in the previous chapter. In his writings on the African disappearances, Schüle presented climatic, biogeographical, anatomical, physiological and behavioral arguments of his own to support Paul's thesis.

I found Schüle's behavioral arguments particularly interesting. The actions of non-hominin animals are, he points out, much more rigidly circumscribed by instinct than those of our family. Just as the birds and animals of the Galapagos cannot, he points out, learn to fear large terrestrial carnivores, whales cannot alter the inappropriately trusting behavior they display in the presence of our species. Because we have only begun to hunt them in the last few centuries, whales simply haven't had time to evolve appropriate behavioral reactions to our species. Melville's *Moby Dick*, Schüle tells us, is "ethologisch ein Unding"—an ethological absurdity. In reality, he explains, deaths of whalers in open boats were generally accidental, rather than the results of aggression by the whales they were pursuing.

Because it occurs close to land, *Megaptera novaeangliae*, the humpback whale, has been hunted by our species for several centuries. The population of this animal has, over that time, been reduced to about a third of its pre-whaling level. But the surviving members of this species do not, as one might expect of such intelligent animals, react to humans with fear or aggression. Consider this meeting between a humpback and a kayaker off the northwest coast of Hawaii's Big Island:

> Gently the whale rose in the water, until it loomed like a gleaming gray rock, and I could see its huge eye, just below the surface, checking me out. Its gaze seemed cool, but not unfriendly; more curious than anything else. It seemed to be wondering what something so small was doing "way out here." Its enormous pectoral fin extended under my kayak as I floated like a gnat beside it, and it was impossible not to consider the fact that it could easily swat me like that same gnat, had it wished. But the gentle smoothness of its movements, compared to the wildly playful breaches and leaps it had been executing a short while earlier, indicated an understanding of my fragility, and its unwillingness to do me harm.

In April of 2001 my son Eric paddled a surfboard up to a southern right whale lying off the middle beach at Plettenberg Bay, on South Africa's southern coast, approaching to within fifteen feet of the animal's head. The whale didn't react at all. Intelligent as many of them are, whales don't have the cognitive power to "figure out" that our species can be very dangerous to them, or the linguistic powers to communicate that kind of realization to other members of their respective species. Like all non-human animals, they must evolve appropriate instinctual responses rather than "figuring out" how to react, and a few centuries of persecution by our species simply hasn't provided a long enough time span for that evolution to take place.

Elephants have, on the other hand, been hunted by hominins for many hundreds of thousands of years. It's not surprising, therefore, that it would be impossible at best, and suicidal at worst, to approach a wild elephant as closely as the Hawaiian kayaker, and my son, got to their whales. In areas where they're well-protected and frequently visited by humans, elephants can stay relaxed in the presence of people inside motor vehicles, but even in those circumstances, they frequently become nervous or irritable. Drive the Kruger Park's 200-mile length from Malelane up to Pafuri, and, even if you don't approach them closely, you're likely to provoke some degree of annoyance in the elephants you'll meet along the way: head-tossing, trunk-raising, ear-flaring, or assertive approaches that will make you stop or retreat. You'll be fairly safe, though—attacks on vehicles happen very seldom. When they do take place, however, they can be terrifying. George Wittemyer, working, as it happened, on the Save the Elephants project, was lucky to survive the total destruction of his 4x4 truck by a bull elephant in Kenya's Samburu National Park in May of 2001:

> [H]e hit us so hard that the car was lifted into the air, and flipped once again onto its wheels. The force of that blow was tremendous—beyond anything I could fathom an animal was capable of. I could see him clearly at this point, coming for my door. I felt he would not quit until he had finished us.

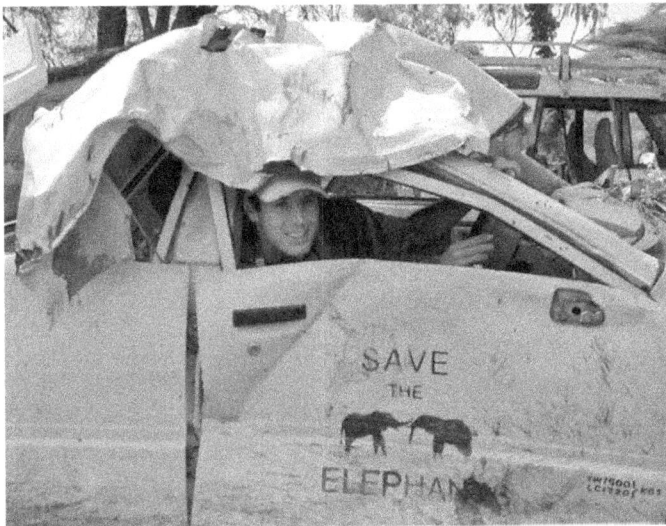

Move around in Africa's wilderness areas on foot, and the risk of a violent confrontation with an elephant increases considerably. In the one-year period preceding February of 2005, five elephant attacks on humans took place in South Africa, involving two walking parties protected by armed escorts, an unarmed pedestrian, and a ranger riding a bicycle. Those attacks resulted in the deaths of three humans and at least three elephants. In April of 2005, a tourist walking in the bush was killed by an elephant in Uganda's Murchison Falls National Park.

I won't overstate this argument by denying that elephants can become completely relaxed around humans. One particular herd of wild elephants—a group that spends a lot of time in the unfenced staff village at Skukuza in the Kruger National Park—has become accepting of humans walking very close to them. Situations such as these cannot, however, disprove the existence of a strong tendency among elephants that are not habituated to close human contact, to feel fear and/or irritation when a member of our species approaches them.

In the 1990 *Homo* article I mentioned earlier ("Human evolution, animal behavior, and quaternary extinctions: A paleo-ecology of hunting"), Schüle suggests that defensive evolution added an extraordinary level of intelligence to the instinctual armamentarium of the surviving elephant species:

> Surviving terrestric [sic.] megaherbivores live and evolved where man also evolved. Supposing that early hominids hunted the ancestors of living elephants, one may postulate a reciprocal influence over the evolution of behavioural patterns in both. Today's elephants are the most cerebralized of the ungulates, and have a complicated social structure. In their herds not only the mothers but also any other member of the herd will attack and even kill anyone they consider dangerous to their young, including humans.

Elephants have not, of course, become smart enough to emulate the inventiveness that has become a way of life for our species, but their intelligence has, nonetheless, added a great deal of flexibility to their behavior. They help each other during birth, injury or sickness, and females have been seen to adopt orphaned members of their species—a very unusual phenomenon in the animal world. Elephants pick up bones that belonged to members of their species (particularly members they have known) and handle them obsessively, and they cover their dead infants with branches—behaviors that speak of an awareness going well beyond that of any other ungulate.

Earth's other surviving megaherbivore species don't share this exceptional level of intelligence, but they are, like the two still-existing elephant species, equipped with strong instinctive feelings of fear and aggression toward humans. Black rhinos are notoriously "ill-tempered" in the presence of our species, and extraordinarily fearful of humans. Fortunately for both them and us, that fear is usually stronger than their aggressive impulses toward us.

Hippopotamuses kill more humans than any of Africa's other wild animals. "Provoked," Chris and Stuart Tilde tell us in their *Field Guide to the Mammals of Southern Africa,* "the hippopotamus can be extremely dangerous..." In areas where they see relatively few humans, the mere sight of a member of our species standing on the banks of one of their pools is often all the provocation it takes to get them going. I've had to beat several hasty retreats from these animals while fishing in Botswana and Zimbabwe.

Pliocene and early-Pleistocene megaherbivores would not, Schüle reasoned in his 1990 "Paleo-ecology of Hunting" article, have been equipped with this instinctive aversion to our species, nor with the exceptional intelligence of the surviving elephant species:

Upper Tertiary and Quaternary megaherbivores were in exactly the same positions as today's whales: a new predator had appeared and hunted animals never hunted before. For the whales,
time is too short to develop new patterns of behaviour. Time
was too short, too, for most Quaternary megaherbivores as well.
Because of the megaherbivores [sic.] lack of fear-patterns, the
problem for early hominids was not to get close to them, but to
find a means to kill the megaherbivores.

CHAPTER 5

Europe with Some Remarks About Asia

WHEN MEMBERS OF *HOMO SAPIENS* LEFT AFRICA AROUND 80,000 years ago they encountered the still-existing *Rhinoceros unicornis*, the one-horned Indian rhinoceros, as well as two smaller species that are now, in the early years of the twenty-first century, on the brink of extinction: *Rhinoceros sondaicus*, the one-horned Javan rhino (a smaller variant of the Indian rhino), and *Dicerorhinus sumatrensis*, the small, hairy, two-horned Sumatran rhino. They also got to know four rhino species that *have* become extinct. The largest of these was a huge grass-eating rhino called *Elasmotherium sibiricum* which lived in and around what is now Russia. Even bigger than the white rhinoceros, *Elasmotherium* carried a single, enormous horn whose base extended from its nose to its forehead. The other three extinct rhinos were, respectively, two woodland Eurasian species in the genus *Stephanorhinus*, and a steppe species called *Coelodonta antiquitatis*. *Coelodonta*, the so-called "woolly" rhino, was a grass-eater similar to, but a little smaller than, the white rhino. It occupied an enormous range stretching from China to Spain.

Between 17,000 and 13,000 years ago, a human artist created this image of a *Coelodonta* rhino (in the bottom left of this photograph of the so-called "Dead Man's Gallery" of Lascaux cave in southern France):

To discuss the meaning of this image we have to talk, briefly, about the rhino language of sex and dominance.

Only male rhinos who've established dominance in a particular territory are allowed to spray clouds of aerosol urine out behind them. (A rhino's penis points backward when it isn't erect.) Those urine mists swirl over the dominant male's hindquarters and over the surrounding vegetation to announce his status and, therefore, his sexual eligibility. They convey a message that is of the highest importance in rhino society. Subordinate males must urinate in a "non-aerosol" stream straight down onto the ground to show that they have no intention of usurping the sexual privileges of the "owner" on whose property they're living. If they don't urinate in this modest manner, the dominant animal will chase them off, injuring or killing them if they resist.

Defecation is an even more complicated matter than urination in rhino society. Rhino dung heaps or middens are located, Richard Estes's *Safari Companion* tells us, "like signposts along rhino paths and territorial boundaries." The dominant male—and *only* the dominant male—has the right to perform the dung-spreading ceremony while he's defecating. In this centerpiece of the socio-sexual lives of rhinos, he kicks a hind leg back, and drags it through the dung midden in a gesture as formal and stylized as the *battement tendu* in which a human ballet dancer draws his or her toe across the stage floor. This *battement tendu* is such a fundamental part of the rhino language of sex and dominance that males start practicing it as small calves. The dominant bull raises his tail and defecates explosively while he's doing this, ejecting softball-sized dung-spheres onto the midden in a compelling blend of drama, beauty and scatological humor. (The use of dung heaps as territorial and sexual notifications runs deep in the rhino family and even extends to some of its surviving relatives in the odd-toed or "perissodactyl" order, horses and tapirs.)

The late Björn Kurtén, one of the greatest authorities on Pleistocene mammals, found the Lascaux rhino image puzzling: "What," he asks on page 125 of *The Ice Age*, "is the meaning of... the six black dots?" In his *The Cave of Lascaux—the Final Photographs*, Mario Ruspoli says they are "a sign meaning the end."

Coelodonta was a close relative of the two African rhino species, and it seems clear that the Lascaux rhino, with its right back leg trailing in the *battement tendu* position and its raised tail, is performing the dung-spreading ceremony. The six black dots are his dung balls. The fact that no one else has, to my knowledge, picked up on this interpretation of the rhino painting in Lascaux

Cave is an indication of how few present-day members of our species have the opportunity of observing in the wild an animal that was once such a familiar part of human life.

* * *

Fifteen thousand years ago, all humans grew up surrounded by animals, birds and fishes. Children living in Europe at this time could probably distinguish mouse, vole, and shrew species with an expertise that only a handful of specialist zoologists possess today. Migrations of reindeer, salmon, geese and swans would have marked the passage of their seasons. Those children would have watched great auks extending their wings for balance as they waddled across wave-washed rocks on both the Atlantic and Mediterranean shores. (Great auks are depicted in this position in paintings in the recently discovered Cosquer Cave near Marseille.)

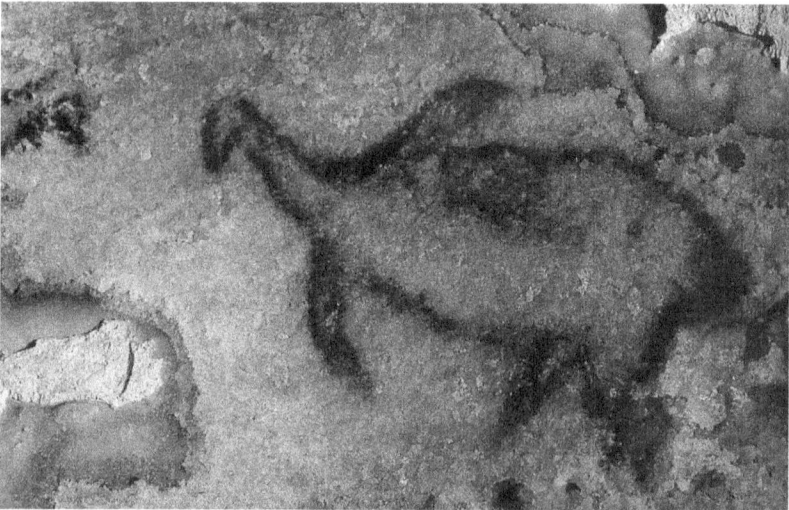

And, no matter where in Europe those ice-age children lived, the moaning of lions and the whooping of hyenas would have been as familiar to them as emergency vehicle sirens and car alarms are to their present-day counterparts. The presence of those lions and hyenas—and that of leopards, brown bears and wolves—wouldn't have soured their day-to-day lives with fear, but it would have kept them focused on the danger of wandering away from their families.

Many of the mammals that lived in the Europe of 15,000 years ago are still found there. Lynxes, beavers, wild cats, otters, badgers and other mustelids, bats, rodents and shrews of various kinds, foxes, wolves, bears, horses, cattle (called "aurochs" in their extinct wild form), ibex, chamois, reindeer, roe deer, fallow deer, red deer, moose (called "elk" in Europe), and bison ("wisent") are all in this category. Some of these survivors (like bison and wolves) hang on in pitifully small numbers, while the cattle population has, like that of humans, become hugely inflated.

Other mammals have disappeared from Europe but are still found elsewhere—lions, hyenas, leopards, saiga antelopes, and musk oxen are in this category. Still others have disappeared completely: the giant deer *Megaloceros*, a large vegetarian bear called *Ursus spelaeus*, the *Coelodonta* rhino, and the mammoth are in this category.

Mammoths are often lumped together with dinosaurs as "prehistoric" animals, but dinosaurs became extinct 65 million years ago, long before there were any big mammals at all, while mammoths were thoroughly modern members of the elephant family that only evolved about 6–8 million years ago in response to the same phenomenon that gave rise to humans and white rhinoceroses—the Miocene drying-trend that stimulated the development and spread of grasslands. Grass is an excellent and abundant source of nutrition, but it's extremely tough and fibrous, and protected, in addition, with abrasive silica. Mammals have evolved two main strategies for processing it. One involves the development of a specialized digestive tract populated with bacteria that have become increasingly good at breaking down its fibers chemically. The other is the evolution of tough, specialized teeth that can break those fibers down mechanically.

Mammoths, and their close relatives Indian and African elephants, chose the latter path. The ridges or "lophs" on their molars and premolars increased in number, and became folded together, resulting, eventually, in chewing teeth consisting of a tightly packed series of buried enamel sheets that filled the tooth from top to bottom.

The complexity of the new molars made them so big that there was only room for a few of them in the jaw. Gomphotheres had already reduced their relatively simple "tetraloph" cheek teeth to twelve, and mastodonts went down to ten. The Elephantidae (the suborder to which mammoths and the three surviving elephant species belong) got down to eight cheek teeth—a single "main grinder" doing the bulk of the work in each quadrant of the mouth, together with a second one "waiting in the wings" to replace it by moving in from the back of the jaw.

The production of two sets of teeth—"milk" or "deciduous" teeth followed by an adult or "permanent" set—is such a common and widespread characteristic of placental mammals that it's thought to be an ancestral condition. Like their distant relatives, manatees, elephants have, however, departed from it—their cheek teeth are replaced no less than five times. The first two teeth to appear are premolars. The remaining three are molars. All teeth other than molars and premolars have disappeared in the Elephantidae, with the exception of two upper incisors that are elongated into tusks which could be over ten feet long in African elephants, and up to sixteen feet long in the largest of the mammoth species, the Columbian. Those two incisors, plus two cheek teeth at a time in each quadrant of its mouth, give the Elephantidae its total of ten teeth at a time.

The living members of the Elephantidae give us a very good idea of how mammoths must have conducted their social lives and raised their offspring. The two still-living elephant species show us, too, how an animal the shape and size of a mammoth or a mastodont must have *moved*—what its living appearance might have been.

Hoping to get a similar clue to the movement and appearance of the giant *Megaloceros* deer we spoke about a few paragraphs back, I followed a group of eland that was moving along the opposite bank of a ravine in Suikerbosrand Nature Reserve near Johannesburg. Moose, which top out at around 1,750 pounds, are currently the biggest deer in the world, but the eland, a giant antelope, can weigh up to 2,000 pounds. I wasn't as impressed as I was expecting to be, by the size of those eland, until they were joined by what looked like a herd of miniature zebra. Only then did things shift into perspective: the "miniature" zebras were, of course, normal-sized animals, and the eland were suddenly revealed for the unlikely giants that they really are. Imposing as they are, eland still weigh only about 70 percent as much as *Megaloceros* did. Seven feet tall at the shoulder, the giant deer carried antlers twelve feet wide, and weighed somewhere between 2,500 and 3,000 pounds—as much as a big giraffe. Because the top speed of big animals tends to be absolutely faster, but relatively slower than that of small ones (the naval architect William Froude described the same phenomenon in relation to ships and, indeed, waves), a running *Megaloceros* must have had the intriguing slow-motion impression created by running elands or giraffes.

* * *

Europe's animal inhabitants divided that continent among them on a "timeshare" basis. During the glacial cycles, the continent was occupied, for the most part, by cold-adapted animals like mammoths, woolly rhinos, and reindeer. During the warm interglacial intervals this "cold-weather fauna" retreated to the northern reaches of Eurasia, while Europe's southerly regions passed into the possession of warmth-loving animals like the straight-tusked elephant *Paleoloxodon*, the woodland rhinos *Stephanorhinus hemitoechus* and *S. kirchbergensis*, and the hippopotamus species that still lives in Africa. Throughout the last million years, these "warmth-loving" animals had been driven southward many times by the glacial episodes that have been occurring every 100,000 years on average. These glaciations either pushed them right out of Europe to find refuge in Turkey or the Middle East, or restricted them to Europe's Iberian, Italian and Balkan peninsulas. After each cold episode, they would return to Northern Europe, as warm, moist interglacial conditions spread broad-leaved forests back into that region.

The warmth-adapted megafauna did not, however, return after the last glaciation, because all of them (except *Hippopotamus amphibius* which survived in Africa) had become extinct by about 25,000 years ago. *Paleoloxodon* was already gone by about 50,000 years ago. It's highly unlikely that the latest glaciation killed these animals off. They had, as we've just seen, successfully survived all the previous Pleistocene glaciations—some more severe than the last one.

* * *

Europe's extinct megafauna included a member of the human family. That species had evolved from beings very similar to *Homo erectus*. These *erectus*-like beings, which are also referred to as *H. heidelbergensis*, probably continued to exchange genes with their Asian and/or African relatives from time to time. By the height of the great Riss glaciation some 150,000 years ago, they had, however, evolved a body shape distinctive enough to differentiate them from their immediate ancestors. In 1869 William King named the new species *H. neanderthalensis*.

Heidelbergensis-neanderthalensis had, as we'll see in Chapter 12, been expert and regular hunters of big game right from the time they entered Europe. They also knew how to make fire, and used their stone tools to fashion sophisticated spears and other implements out of wood. Between 350,000 and 410,000 years ago, at Bilzing-sleben in east-central Germany, a *heidelbergensis* individual incised a pleasing, abstract pattern onto a bone artifact, whose discovery made the chief researcher at that location, Dietrich Mania, suspect what he calls "the development of a capacity for abstract thought, and the presence of language" ("...die Herausbildung der Fähigkeit zum abstrakten denken, und das Vorhandensein einer Sprache").

The discovery of the incised patterns constitutes a particularly significant enlargement of our understanding of *heidelbergensis*, because the production of decoration and art is regarded—even more jealously than some people regard the use of bone tools and the ability to hunt big game—as an exclusive province of *Homo sapiens*. Although there is no evidence of any kind that the *heidelbergensis-neanderthalensis* line produced representational art, those species may not have been indifferent to decoration and adornment: ochre and manganese oxide pigments have been found at several sites associated with *heidelbergensis*, while ochre and ochre-grinding paraphernalia have been found at sites occupied by Neanderthals.

Our kind of human, *Homo sapiens*, first entered Europe between 45,000 and 50,000 years ago, probably via what is now Bulgaria. By some 38,000 years ago, *sapiens* had reached present-day Portugal. A second wave of *sapiens*, equipped with the so-called "Gravettian" complex of technologies, entered Europe some 30,000 years ago from what is now Russia. Europe hadn't been occupied by hominins for as long as southern Asia when our kind of human made its first appearance there. The mammoths, woolly rhinos, giant deer and lions that modern humans recorded in their cave paintings and carvings would not, therefore, have had the same opportunities to adapt to a hominin presence that preserved much of Africa's mega-fauna as well as South Asia's elephants, rhinos and big cats.

As I mentioned in the previous chapter, diversity of South Asia's biggest animals experienced an early extinction episode that coincided with the decrease in Africa's big-animal diversity around 1.4 million years ago. South Asia lost proboscidean species in that episode, as well as one or more *Hexaprotodon* forest-hippos. At about the same time, South Asia's *Megantereon* sabertooths followed these big herbivores into extinction. South Asian megafaunal losses would, like those suffered by Africa, continue at a reduced level throughout the Pleistocene. Later in that epoch, South Asia would lose *Gigantopithecus*, the largest ape that has ever existed, and the giraffe/okapi *Sivatherium. Pachycrocuta*, the giant hyena, survived longer than *Megantereon* in this region, but it would also slip into extinction long before the end of the Pleistocene.

The final spasm of megafaunal extinctions to hit Southern Asia took place after *Homo sapiens* evolved in Africa and moved into that region some 80,000 years ago. That spasm was approximately contemporaneous with the large-scale extinction of the megafauna of the New World, North Asia and Europe, and with Africa's final, and relatively small, extinction-spasm that we discussed in Chapter 3. In this end-Pleistocene spasm, South Asia would lose the last of its mastodonts, and the last of its machairodonts, the scimitar-tooth *Homotherium ultimum*. Like Europe, it would also lose a hominin species: *Homo erectus*, which had made its first appearance in South Asia nearly 2 million years ago, makes one of its last appearances in that region between 117,000 and 108,000 years ago at Ngadong, Java. *H. florisiensis*, either a dwarf variant of *erectus*, or more likely a late-surviving australopithecine, appears to hang on until about 15,000 years ago.

A strong current of denial runs against the idea that humans could have been responsible for the extermination of the European and other Pleistocene communities of megafauna. "Worldwide," the Australian Museum's website tells us, for instance, "there is no evidence of Indigenous peoples systematically hunting nor over-killing megafauna... For social, spiritual and economic reasons, First Nations peoples harvested game in a sustainable manner." Even a plain-as-day example of extinction caused by pre-industrial humans, like the disappearance of New Zealand's 500-pound moas, giant eagles and dozens of other species within a century after the ancestors of the Maori discovered and settled that island group in the fourteenth century, can attract this of kind of denial: a 2004 study (whose conclusions were later refuted) tried to prove that moa populations were already declining before humans arrived in New Zealand, and that our species could not, therefore, have been "entirely to blame for wiping out the moas..."

CHAPTER 6

Australia and North America

WHILE I WAS WRITING THIS CHAPTER, I HEARD THE PRESENTER of a TV show about Australian wildlife tell his viewers that "before man arrived in Australia, the kangaroo had no enemies."

Nothing could be further from the truth. When the first humans arrived in Australia some 65,000 years ago, kangaroos were, in fact, being preyed on by a marsupial "lion" called *Thylacoleo carnifex*. Until recently, it was thought that *Thylacoleo* was a leopard-sized animal, but evidence marshaled by Stephen Wroe of the University of New South Wales suggests that it might have been closer in size to a female lion or tiger. Barring the discovery of a painting of *Thylacoleo*, we'll never know what kind of markings it had.

Thylacoleo was more of a "rat-bear" than a cat—the animal's body was stocky and muscular rather than lithe. Two big incisor teeth pressed against each other in a rodent-like way in the front of each of its jaws to form a kind of "beak" for seizing its prey. Behind the remaining incisors—which were small—lay one huge premolar in each quadrant of the mouth. These four teeth had developed into long carnassials or "meat cutting" blades. The front cnds of the upper carnassials ended in a pair of low "summits" that could, at a quick glance, look like the canine teeth which were absent or undeveloped in this animal. The teeth situated behind the carnassials were also small. The toes of this big marsupial predator were equipped with claws, two of which, situated on powerful opposing "thumbs," were retractable:

Reproduced with kind permission of Peter Schouten, wildlife artist and creator of this image.

Thylacoleo may not have been particularly cat-like, but the wolf-sized marsupial predator *Thylacinus cynocephalus* was astonishingly and faithfully dog-like. Named "Tasmanian wolf" because of this resemblance, *Thylacinus* was only "Tasmanian" in the sense that a somewhat dwarfed, coyote-sized version of this animal survived on the island of Tasmania until it was exterminated in the 1930s by the British colonists who had wiped out the island's original human inhabitants during the previous century. *Thylacinus* had, along with the "Tasmanian devil" *Sarcophilus harrisii*, become extinct on the Australian mainland about 3,500 years ago, soon after the dingo, a placental or "real" dog, was brought to Australia by humans long after the first wave of human immigration. By that time, however, Tasmania, which had, like New Guinea, been joined to Pleistocene Australia, had already been cut off from Australia by the post-glacial rise in the sea level, so the dingo (whose competition *Thylacinus* could not, apparently, survive) didn't make it there.

Film footage of *Thylacinus*, taken in a zoo in the early thirties, shows an animal that looked like, and trotted like, a dog. Given the fact that "real" wolves are much more closely related to sheep (or to whales or bats, for that matter) than they are to this marsupial "wolf," the shape and arrangement of their forty-two teeth is remarkably similar to that of *Thylacinus*'s forty-six. Among the few obvious details that betrayed the fact that *Thylacinus* was completely unrelated to the dog family to which it bore such a remarkable resemblance, were a tail that was thicker at its base than the tails of "real" dogs, and jaws that could open much wider. *Thylacinus*'s "pups" were carried in a backward-facing pouch. The hind end of the animal's back, including the thick base of its tail, was covered with about twenty dark, parallel stripes. Because of these stripes, some people have taken to referring to *Thylacinus* as the Tasmanian "tiger," but that name obscures the animal's remarkable evolutionary convergence with the dog family.

The extermination of *Thylacinus* has left only three small marsupial predators in Australia: the terrier-sized Tasmanian devil and several mongoose-like "quolls" in the genus *Dasyures*.

* * *

Australia's extinct "wolves," "lions" (and a ninety-pound carnivorous "rat-kangaroo" called *Proleopus oscillans*) represented only that subcontinent's medium-sized predators. Its largest flesh-eaters were all reptiles. One of the most remarkable of these was a huge monitor lizard called *Megalania prisca*. *Megalania* grew to just over twice the length of its surviving relative, the "Komodo dragon" which can grow up to ten feet long, and inhabits a few Indonesian islands near Australia's northern coast.

Other big reptilian predators living in Australia when humans first arrived there were the outsized "python" *Wonambi* and a long-legged crocodile called *Quinkana*. The length of the latter's legs, and the fact that its tail was rounded in cross section rather than flat-sided like the tails of water-dwelling crocodiles, suggest that *Quinkana* lived and hunted on land. Australia also had several species of water-dwelling crocodile of which two have survived. The larger of these two survivors, *Crocodylus porosus*, which occasionally reaches eighteen feet in length, is presently the world's largest reptile.

These big reptilian predators would certainly have killed kangaroos, but they would also have focused their efforts on larger prey, like Australia's marsupial "rhinos." Those "rhinos" were members of the "Diprotodonid" group, one of whose genera, *Diprotodon*, consisted of two or more species that were as big or bigger than a black rhinoceros. Australia's biggest predators would also have targeted a buffalo-sized marsupial called *Zygomaturus*, and the strange *Palorchestes*, a one-ton marsupial "ground sloth" that could balance on its powerful tail and hind limbs like a kangaroo while reaching up with the huge, curved claws on its forelimbs to pull trees and branches into the reach of what was possibly a short, tapir-like trunk.

The mammalian and reptilian predators we've been talking about would also have preyed on seven species of "giant short-faced" kangaroos that disappeared along with the "rhino," "buffalo," and "ground-sloth" analogs I've just mentioned. Eight large species of "ordinary" kangaroos, and five large wombat species, were also exterminated or dwarfed.

Add to this fanciful collection of extinct giants a family of horned tortoises in the genus *Meiolania*, whose largest species reached a weight of 500 pounds. Include with them, too, a list of extinct birds that contains, inter alia, two flamingo species, a pelican, one or more giant eagle species, flightless cuckoos, flightless pilot birds, a giant "mallee fowl" and the 450-pound *Genyornis newtoni* "ostrich," and you can begin to get an idea of the enormous impoverishment the Australian fauna has experienced in "near time"—that is, since the arrival of humans in their homeland some 65,000 years ago.

* * *

In the year 2001, a study initiated by Tim Flannery and Richard Roberts, and based on a dating technique known as "optically stimulated luminescence," singled out the millennium between 46,000 and 47,000 years ago as the time when many species of Australian megafauna died out abruptly over a wide area of that island-continent. These findings are in broad agreement with the results of a 1999 study by Gifford Miller of the University of Colorado, who used another dating technique—amino acid racemization—to establish that the youngest egg shell fragments of the giant *Genyornis* "ostrich" are between 45,000 and 55,000 years old.

Fire played an expanded role in the ecology of Australia after humans first arrived there some 65,000 years ago. Whether humans do so by accident, or to signal other groups, or to help their hunting, or stimulate fresh plant growth, they start plenty of fires. The result was that, relatively soon after the arrival of our species in Northern Australia, dry forests dominated by Araucaria trees were supplanted by plant communities dominated by eucalypts, which "rely" on fire to help them compete against plants that are more vulnerable to it than they are. Using sophisticated "pre-treatment" strategies for removing contaminants in his samples, Chris Turney of Queen's University in Belfast pushed the horizon of the radiocarbon method back from 40,000 to 60,000 years. In doing so, he was able to establish that an abrupt increase in burning, and an accompanying change in vegetation, had taken place in Queensland sometime between 45,000 and 55,000 years ago.

* * *

While Australia's extinct megafauna disappeared on the edge of the 40,000-year "window" in which radiocarbon dating is effective, the

New World's "Serengetis" of big animals and birds became extinct well inside that window—near the start, to be more precise, of the "Younger Dryas" cold phase some 12,870 "calendar," or 11,200 "radiocarbon" years ago.

Let's talk briefly—before we return to the issue of when North America's megafaunal extinction took place—about why the radiocarbon date I've mentioned in the preceding paragraph is so far out of step with the calendar or "real" date it represents.

Radiocarbon dating is based on the assumption that the amount of C14 in the atmosphere remains constant. That assumption doesn't hold in the real world. Variations in the Earth's magnetic field, and in cosmic background radiation, affect the rate at which C14 is being "manufactured" in the atmosphere. The big, back-and-forth lurches in temperature that took place near the end of the last glaciation muddied the picture further, by causing sudden and relatively large variations in the ratio of C12 to C14 in the atmosphere. Those variations make time, as measured by radiocarbon dating, appear to speed up, slow down, and even run backward for brief intervals. We're dependent, therefore, on data gleaned from lake sediments, annual precipitation layers in ancient ice, tree-ring sequences and other dating methods to produce "calibrated" or "calendar" dates from the radiocarbon results.

All the dates mentioned in this book are, unless I've indicated otherwise, expressed in calibrated or calendar rather than in radiocarbon years.

* * *

Although the New World's "Serengetis" disappeared, as we've just seen, well within the radiocarbon dating window, many of their members lack the kind of radiometric records that are needed to calculate the dates of their last appearances with any degree of precision. Reliable radiometric histories are only available for some fifteen of the approximately thirty-five genera of mammals weighing over 100 pounds, which North America lost near the end of the Pleistocene. Those measurements tell us that those fifteen genera all became extinct between 13,000 and 12,400 calendar years BP.

Some of the remaining twenty genera may have become extinct before or after that 600-year-long spate of more or less securely-dated extinctions, but the picture that emerges is, nevertheless, one in which North America loses a megafauna that was much richer than that of present-day Africa in something like a thousand years—a millisecond of geological time.

The fact that somewhere between 30 and 70 million bison were living on the prairies of North America in the early nineteenth century demonstrates the fact of this catastrophic impoverishment rather than casting doubt upon it: *Bison bison* could never have become as plentiful as it was in the 1840s while it was still sharing the prairies of the "American Serengeti" with a dozen other large herbivore species (including its relative *Bison latifrons*). The real Serengeti contains, for instance, a great many wildebeest (1.4 million of them still live there), but the entire Africa-wide population of the blue wildebeest (the species that lives on the Serengeti) could never have come near to the nineteenth-century abundance of *Bison bison*. This is so because blue wildebeests have always had to share the resources of the African veld with relatives like black wildebeests, hartebeests, topis, tsessebes and blesboks, and a host of unrelated grass-eaters ranging from elephants through white rhinoceroses to oryxes.

The population of the one surviving American bison species, artificially boosted by the extermination of most of America's big animals between 13,000 and 12,000 years ago, was further inflated after the fifteenth century CE when a large proportion of North America's human population was wiped out by the introduction of smallpox and other Eurasian diseases. The enormous bison herds of the mid-nineteenth century were, therefore (in the words of Colin Tudge's *Time before History*), "an artifact—like a rash of ragwort on a bombsite."

* * *

When humans first entered North America, at least one cheetah species, *Miracinonyx trumanii*, was living there. (Two older American cheetah species, *M. studeri* and *M. inexpectata*, had probably drifted into "natural" extinction by that time—that is, extinction not caused by our species.) We can guess that the American cheetahs probably focused their predatory efforts on the pronghorn and forked-horn "antelope" species that were America's equivalent of Africa's gazelles, because the only surviving member of the pronghorn/four-horn group, *Antilocapra americana*, is still one of the fastest-running hoofed animals in the world. Evolution would never have made the enormous investment in the circulatory, respiratory and muscular-skeletal development required for that kind of speed without a pressing reason for doing so.

The cat family originated in Eurasia/Africa, but a member of that family that reached North America in the early Miocene, probably over the newly formed Bering Land Bridge, founded the cheetah-cougar family in its new homeland.

Recent research suggests that the American cheetah *Miracinonyx* may not be related to the Old World cheetah *Acinonyx* but might have evolved in parallel with the latter in North America, but the question hasn't been settled. Morphology or body shape could suggest that Old World cheetah is, in fact, related to the cougar/*Miracinonyx* group. Both *Acinonyx* and cougars have small, rounded heads and long, relatively heavy tails; neither can roar but both chatter in squeaky, bird-like voices. The black "tear marks" running from the inner corners of the African cheetah's eyes down to the corners of its mouth are echoed by prominent black smudges on the same area of the cougar's face. The cheetah's spots, too, resemble those that adorn juvenile cougars: they're solid dots rather than the very different "rosettes" or "clusters of spots" displayed by jaguars, leopards, adolescent lions and many of the smaller cats.

End-Pleistocene North America was inhabited, too, by lions and jaguars. The former species was similar to, or identical with, the lions that still inhabit Africa and a small part of India. For reasons we spoke about elsewhere, both the lions and the jaguars were, however, about 25 percent bigger than the present-day members of their species.

In addition to the big cat species I've just mentioned—which still exist *somewhere* on the planet, if not in North America itself— Pleistocene North America was inhabited by two large felids that have become completely, rather than just locally, extinct: the saber-tooth *Smilodon* and the "scimitar-tooth" *Homotherium*. *Smilodon* was a closely related descendant of the *Megantereon* sabertooth which had lived (along with *Homotherium* and the "dirk-tooth" *Dinofelis*) in Africa until some 1.4 million years ago.

Elsewhere in this book, I argue that sabertooths disappeared from Africa at this early date because hominids had, by that time, begun to dominate the business of hunting large prey effectively enough to shut out other predators that were dependent on such prey. If I'm right about this, the relatively late arrival of hominids in Europe would explain why sabertooth cats survived there long after they had become extinct in Africa. (*Megantereon* only disappeared from Europe some 800,000 years ago, while *Homotherium* made its last appearance as recently as 25,000 years ago, when our kind of human, *Homo sapiens,* was already taking possession of that continent.) The fact that the hominid family— in the person of *Homo sapiens*—only entered the New World near the end of the Pleistocene, would explain, therefore, why *Megantereon* (in its *Smilodon* avatar) and *Homotherium* survived there until that very recent time.

The six big-cat species we've been talking about shared end-Pleistocene North America with four dog-like species: the coyote, the "ordinary" wolf, a more heavily-built "dire" wolf, and the dhole—a wild dog that is still found in Asia.

The still-surviving black bear also lived in North America when humans arrived there, as did the grizzly bear. Although it's considerably larger than a lion, the grizzly wasn't the biggest predator to inhabit the New World when our species first arrived there: that position was held by the giant short-faced bear *Arctodus*, which was probably the largest land-dwelling carnivorous mammal ever to exist. "Awesome" has become an overused word of late, but *Arctodus*, equipped with long legs adapted to running pursuit, must truly have been an awesome animal. A fourth bear species, *Tremarctos floridanus*, was living in North America when humans first arrived there. *Floridanus* was closely allied to (but considerably larger than) *Tremarctos ornatus*, the mainly vegetarian "spectacled bear" which survives in South America.

Another predator living in end-Pleistocene North America, was the "cursorial" or running hyena *Chasmaporthetes* which disappeared from Africa at about the same time that the sabertooths did—about 1.4 million years ago.

* * *

Twenty thousand years ago, this huge assortment of big carnivores was living off a much larger assortment of big herbivores. All the still-existing North American species such as moose, elk, caribou, white-tail and mule deer, musk ox, bison and wild sheep and goats were present of course, but there were, in addition, the two other species of bison I mentioned a few paragraphs back; a woodland musk ox named *Bootherium*; a "shrub-ox" called *Euceratherium*; an extinct relative of the still-living mountain goat *Oreamnos*; a "stilt-legged" deer called *Sangamona*; a medium-sized one called *Navahoceros*; a caribou-like deer named *Torontoceros*; a moose-like deer called *Cervalces* and a very large moose-variant named *Bretzia*. Unglaciated pieces of North America's northern end were inhabited, too, by the saiga, an antelope that still survives in Eurasia.

If we could use a time machine to travel to the North America of this time, we'd be able to see, too, the varied suite of "antilocaprids"—that is, the pronghorn and forked-horn antelope species that I mentioned in connection with America's cheetahs. Just as present-day visitors to East Africa's game parks might argue about whether they're watching Thompson's gazelle, Grant's gazelle, Speke's gazelle, Roberts's gazelle or Rainey's gazelle, we might, on such a time-trip, get into lively discussions about whether we were looking at Matthew's pronghorn (*Capromeryx furcifer*), the little pronghorn (*C. minor*), the Mexican little pronghorn (*C. mexicana*), Shuler's pronghorn (*Tetrameryx shuleri*), Mooser's pronghorn (*T. mooseri*), Conklin's pronghorn, (*Stockoceros conklingi*) or Quentin's pronghorn (*S. onusrosagris*).

We could get into the same kind of arguments about the camel-llama suborder *Tylopoda* that was represented in its North American homeland by three genera at this time: the genus *Camelops*, represented by perhaps three species, and two llama-like genera called *Paleolama* and *Hemiauchenia*, which may also have contained multiple species.

Three peccary or "New World pig" species still live in the Americas: the collared peccary (or "javelina"), the white-lipped peccary, and the recently discovered Chacoan peccary *Catagonus* (which was first identified by late-Pleistocene sub-fossils in the thirties). During the late Pleistocene, there were two additional species: a long-nosed, long-legged peccary called *Mylohyus*, and a large, "flat-headed" species called *Platygonus compressus*. *Platygonus*'s remains have been found at about 130 sites throughout North America, and it may well have been that continent's commonest medium-sized mammal.

Peccaries and cheetahs were "naturalized" Americans—that is, animals that had come to North America early in their evolutionary history, but the camel-llama suborder *Tylopoda*, the "pronghorn" or antilocaprid family, and the *Perissodactyl* order (i.e. horses, tapirs, rhinoceroses and chalicotheres) were North American born and bred.

Rhinoceroses disappeared from their North American homeland about 5 million years ago. The disappearance of this and a great many other animal groups from North America during the late Miocene was caused by the drop in the planet's average temperature which had been taking place throughout the latter part of that epoch, and accelerated about 5 million years ago. North America was, even after that temperature drop, still a warmer continent than it is today, but its rainfall dropped sufficiently to stimulate the evolution and spread of grasses, including tough, fibrous "C4" species.

This cooling and drying trend, and the accompanying rise of grass-lands, was a worldwide phenomenon, but it seems to have brought about a particularly large number of extinctions in North America. That might have resulted from the fact that North America lacked the relatively warm, relatively moist tropical refuges retained by Africa, Eurasia and South America. (North America had not yet become joined to South America when this Miocene cooling took place.)

The early Miocene saw the ending, too, of a relatively long period of isolation for North America, by starting the periodic openings of the Bering Land Bridge which have been going on ever since that time. This let in successive waves of Eurasian invaders such as probos-cideans, bovids, cervids and felids that may have compounded the difficulties that climatic and vegetational change was causing for North America's "autochthons" or native species.

Long before the North American rhinoceroses became extinct, at least one of its members had dispersed into the Old World (where

rhinoceroses have, of course, survived into the present). That
"emigration" took place in the Oligocene Epoch some 30 million
years ago. North America may still have been connected to Scan-
dinavia via Greenland at that time. Rhinoceroses could, therefore,
have used a "Greenland bridge" to get to Eurasia and beyond. During
the late Pleistocene, three species of tapir lived in North America—
Tapiris veroensis, T. copei and *T. californicus*. Four members of this
genus still exist: three in South and Central America and one in
Malaysia (which it reached via the Bering bridge).

Of the twelve genera of equids, or horse-like animals, that had been
in existence at the beginning of the Miocene, only three survived to
the end of that epoch. North America's mammalian diversity started
growing again, however, in the later Pliocene and the Pleistocene.
During that time, one of the "horse-like" genera that had survived
the Miocene extinction, *Equus*, evolved a rich diversity of species
in North America. By the end of the Pleistocene, *Equus* was—along
with the two other surviving genera, *Hippidion* and *Onohippidium*—
also present in South America.

We can divide the still-existing members of the genus *Equus* into:

1. Caballines or "true horses," which includes the domestic
    horse *Equus caballus* and a wild species called *E. pzewalskii*;

2. Hemionines or Asiatic wild asses which include Kiangs,
    Onagers, Kulans and Khurs;

3. Asinids (the Nubian wild ass *Equus asinus*, and the domesti-
    cated donkey derived from it, as well as the Somali wild ass
    E. a. somalicus); and

4. Zebrines, which include the large, mule-like Grevy's zebra,
    Burchell's zebra (divided into three subspecies including
    the partly striped, extinct quagga), and the ass-like moun-
    tain zebra (represented by two subspecies).

Greg McDonald, a vertebrate paleontologist with US National Park Service, thinks that the "horse" species abundantly represented in the Hagerman fossil beds in Idaho where he is stationed, *Equus simplicidens*, was a close relative of Grevy's zebra. It's widely assumed, too, that the smallest of the American "horses," *E. tau*, was a "pigmy onager," that *E. calobates* was a "stilt-legged onager," and that *E. conversidens* (whose human-butchered remains have been discovered in association with Clovis spearpoints on the drained bed of St. Mary Reservoir in southern Alberta) was a kind of ass.

I'm fully persuaded that ass-like, zebra-like, onager-like and/ or horse-like characteristics were scattered among the New World's diverse equine species. I would have had to have been guided, however, through a "multivariate" analysis of the relevant specimens by someone like my late friend James Brink of the Florisbad Quaternary Institute in Bloemfontein, who had a special interest in equine paleontology, before I could presume to affirm or deny that one or another of the American equine species was identical or close to a particular, still-living member of the genus *Equus*.

How many horse-like species were there in the New World at the end of the Pleistocene? FaunMap, an electronic database documenting the late Quaternary distribution of mammals in the United States developed at the Illinois State Museum and directed by Russel Graham and Ernest Lundelius, lists seventeen North American equines. (Let me add, uncritically, though, that the creators of FaunMap, who also list five North American mammoth species, espouse the "splitting," as opposed to the "lumping," end of taxonomic thinking.)

Despite its wealth of species, the entire *Equus* genus disappeared from the New World at the end of the Pleistocene, along with the South American equids *Hippidion* and *Onohippidium*. The horses later used by American Indians were, therefore, the descendants

of animals imported from Europe. Horses were, with that impor-
tation, completing a journey around the planet by returning to the
continent on which they'd originally evolved.

Equid species had been spreading into Eurasia (via the regularly
opening and closing Bering Land Bridge) since this family first arose
in North America early in the Age of Mammals some 50 million years
ago. The next-to-last wave of equid emigration washed into the Old
World with the arrival of three-toed horses or "hipparions" in the late
Miocene some 10 or 12 million years ago reaching not only Eurasia
but also Africa. Together with the remains of antelope species
found in association with it, the cut-marked femur of a hipparion,
unearthed at Bouri in Ethiopia in the late 1990s, constitutes proof
that the human-like beings we'll meet in Chapters 10 and 11 were
using stone tools for the purpose of butchery near the Pliocene-Pleistocene border, some 2.5 million years ago. As we saw in Chapter 4,
hipparions became extinct in Africa about a million years ago.

We mustn't, of course, imagine that faunal traffic flowed only one
way over the Bering Land Bridge which *Equus* and other American groups used to reach Asia. Many African and Eurasian species
crossed it in the opposite direction to establish themselves in the
New World (and some of those recrossed it thereafter). Old World
immigrants to the New World include bison and other bovids, deer
and cat species, as well as several kinds of elephant. At least seven
elephant-like or "proboscidean" species, divided into three families
(each of which had evolved in Africa), were living in the New World
when our species first entered it.

The mammoth family was represented by at least three species. The
Eurasian woolly mammoth *Mammuthus premigenius* had settled
the northern parts of North America late in the Pleistocene. One or
more waves of mammoths had, however, spilled into North America
near the beginning of the Pleistocene some 1.8 million years ago.
The descendants of these "early arriving" mammoths were, by the
end of that epoch, living in great numbers in the southern, western

and midwestern regions of that continent. Some paleontologists separate these early-arriving or "native" mammoths into two or even three species, *Mammuthus columbi*, *M. imperator* and *M. jeffersoni*, but the current trend is to place them all into the *columbi* taxon.

I'll follow that trend and talk about all America's "native" mammoths as *M. columbi*. We're prepared, after all, to see all African elephants as members of the single species *Loxodonta africana*, even though they're a very variable group. *L. africana* has a "savanna" subspecies, *L. africana africana*, and a somewhat smaller and very different looking "forest" subspecies, *L. a. cyclotes*. Some zoologists recognize the West African pygmy elephant—which is only about six and a half feet tall at the shoulder—as a separate species that they would call *Loxodonta pumilo*, but most would classify it as an even smaller variant of cyclotis. At the other extreme, the desert elephants of the Kaokoveld in Namibia, with a shoulder height of up to thirteen feet, tend to be even taller than "standard" savanna elephants like those found in the Kruger and Serengeti National Parks.

Columbian mammoths had probably evolved a comparable radiation of subspecies by the end of the Pleistocene. These "native" mammoths may have been relatively hairless, or they could have been sparsely haired like Sumatran rhinos. Their tusks could, as we saw in Chapter 4, grow to an astonishing sixteen feet in length. (The longest African elephant tusk on record measured eleven feet, five inches.) Columbian mammoths reached a larger size than both woolly mammoths and African elephants. Big African elephants weigh about 12,000 pounds and the biggest ones reach 15,000 pounds. The biggest Columbian mammoths probably exceeded 20,000 pounds. The smallest Columbian mammoth we know of was, however, even smaller than West Africa's "pygmy" elephant. Assigned to the separate species *Mammuthus exilis* because of this dramatic size difference, this animal had become dwarfed after reaching, and living on, the Channel Islands off what is now Santa Barbara on the coast of Southern California.

The mastodont family had moved out of Africa/Eurasia into North America in the late Miocene some 3.5–4 million years ago. By the time humans reached that continent only one of its members, *Mammut americanus*, was living there. The grinding surface of mastodonts' cheek teeth consisted, as we saw in Chapter 4, of breast-like bumps or cones. Those simple, enamel-covered cones stood in sharp contrast to the closely packed plates of buried enamel that gave the "loxodont" cheek teeth of mammoths their grass-resistant qualities.

While it's clear, therefore, that *M. americanus* had simpler cheek teeth than mammoths did, it may have had a relatively complex liver. Specialist browsers are usually confronted with many more kinds of potentially edible plants than grazers are, and commonly evolve livers that are sophisticated enough to detoxify the wide range of the "defense-poisons" to which this catholic diet exposes them. Livers don't, however, fossilize the way teeth do, so we can only speculate that this may have been the case with *M. americanus*.

Preserved stomach contents confirm, at any rate, that the American mastodont *was* a specialist browser, whose diet included the twigs and leaves or needles of spruce, pine, larch and cedar trees. As one might expect, therefore, *americanus* was the commonest elephant in the forests of what is today the eastern United States. It was, however, widely distributed throughout North America, and its remains have been found as far south as Honduras. Standing about eight feet tall at the shoulder, it was about two-thirds as tall as a Columbian mammoth, but its body was as long or longer than that of the mammoth. Like those of the *premigenius* or wooly mammoth, the bodies of American mastodonts were covered with hair.

The third American elephant family, the gomphotheres, had arrived in North America as long as 15 million years ago, near the middle of the Miocene Epoch. Gomphotheres could not have entered *South* America at this early stage, because that continent would only come into contact with North America just over 3 million years ago. By the time humans arrived in the New World most (if not all) of the members of the gomphothere family had, however, become restricted to South America: *Cuvieronius hyodon* seems to have lived in the colder, more elevated regions of that continent, while *Stegomastodon waringi* and *S. platensis* inhabited its warmer, lowland regions.

The cheek teeth of the gomphotheres were broadly similar to those of the "breast-toothed" mastodont, but they were covered with a somewhat thicker layer of enamel whose surface area was increased by "folds" or "wrinkles." Stable isotope studies tell us that, at least in some areas, gomphothere species combined browsing with grazing. This unspecialized feeding pattern may explain why gomphotheres had, by the end of the Pleistocene, been displaced in North America by mastodonts and mammoths (both, as we've seen, specialist feeders). If humans had somehow disappeared off the face of the earth 15,000 years ago, those two specialists might eventually have displaced the gomphotheres from their South American sanctuary too.

* * *

I've been assuming, so far, that the Columbian mammoth was the largest member of the New World's megafauna. It's possible, though, that Steller's sea-cow *Hydrodamalis gigas*, the supergiant dugong that I mentioned in Chapter 1, could have been even bigger. Estimates of *Hydrodamalis*'s weight range between 12,000 and 24,000 pounds. This species was, as you may recall from Chapter 1, exterminated by humans in 1768—less than thirty years after members of Vitus Bering's expedition discovered it living around the two uninhabited and previously unknown Komandorskiye Islands off the end of the Aleutian chain. That discovery actually took place as a result of Bering and his crew being shipwrecked on the larger of those two islands. Bering was, like many members of his crew, to die on that island, which was later named for him.

Steller's sea cow was very much a member of North America's Pleistocene fauna. The small population discovered in the eighteenth century by Bering's shipwrecked crew was simply the last remnant of a species that had, before human hunting had restricted it to its remote northern sanctuary, extended down to California or Mexico (and down to Japan on the other side of the Pacific). Daryl Domning, an anatomist at Howard University, suggests that its northern "sanctuary" could have been a suboptimal environment for the 1,000–2,000 animals that survived there, and that the giant dugongs might have grown larger in the southerly portions of their range. (*Hydrodamalis* was more closely related to dugongs than manatees—its tail resembled, for instance, the dolphin-like tail of the former, rather than the rounded, paddle-like tail of manatees.)

Georg Wilhelm Steller, the Bering expedition's young medic and naturalist, left us an excellent description of *Hydrodamalis*. Steller's writings make it clear that these huge dugongs would probably have been as accessible to early human hunters as North America's terrestrial megafauna was. They lived in very shallow water and came so close to the shore that Steller and his companions could "hit and reach them with sticks" while standing on dry land. Steller never saw a *Hydrodamalis* submerging its body completely. Members of this species typically lay in the water with their backs protruding, using their curved front limbs (which terminated in "brushes" formed by a dense growth of stiff bristles) to "walk on or pull themselves along" the bottom. Steller also saw them using those front limbs to scrape seaweed from the rocks and hold each other while mating. (He concluded, incidentally, that they were monogamous.)

Like its relative the savanna elephant, Steller's sea cow was a messy, destructive eater: kelp stalks and pieces of other seaweed would float to the surface and wash ashore in large quantities where groups of them were feeding. For some reason, Steller often came upon those feeding parties at places where streams flowed into the ocean. Did these animals need, one wonders, to drink fresh water?

After living on the meat of cormorants and seals for several months, the taste of which they didn't enjoy, Steller and the other survivors of the wreck of Bering's ship devised a way of slaughtering a few of these monsters: thirty to forty men would stand on the shore at high tide, holding one end of a long, stout rope. Another party would row the wrecked ship's "jolly boat" up to the selected animal and thrust a large metal hook, attached to the other end of that rope, into its flesh. The shore party would then start pulling on the rope, while the men in the boat would harass the hooked animal to the point of exhaustion, making it bleed copiously by stabbing it with large knives and bayonets.

While this battle was taking place, the animal's companions would, Steller noted, make unavailing efforts to help it. The shore party would then pull the animal against the shore and hold it there until the falling tide beached its enormous body, enabling them to cut strips of meat and fat off it. "We noticed," Steller tells us, "not without amazement, that a male approached his mate, who was lying on the beach, for two consecutive days, as if to find out what had happened to her."

Steller and his party may not have had the energy needed to build a seaworthy vessel out of the wreckage of their ship and make it back to Kamchatka, if they had not gained access to the palatable, beef-like flesh, and the aromatic oil, of these giants. That oil would, Steller tells us, become "as pleasantly yellow as the best Dutch butter" after it had been left in the sun for a few days. Back on Kamchatka, he and his party spread the word about the Komandorskiye Islands and their tasty "sea cows." As Steller saw it, the animals were present around those islands "...in such large numbers that, in each year, all the inhabitants of Kamchatka's east coast could obtain an abundant supply of meat and fat from them." As we have already seen, however, Steller's prediction vastly underestimated his species' ability to drive a species to extinction.

I can't imagine what the impact of a manatee or dugong ten to twenty times heavier than any of the still-living species might have been. I remember how surprised I was when I saw my first member of this group—an Antillean manatee—surfacing near my boat in a jungle river in South America. I'd vaguely imagined manatees to be as big as large otters, but this animal, ten feet or more in length, and probably over 1,000 pounds in weight, was the size of a beluga whale. *Hydrodamalis,* on the other hand, was the size of a big killer whale. I find myself regretting the fact that my species discovered this marvelous creature—or rather rediscovered it—as early as 1741. We had, after all, developed the awareness required to prevent ourselves from exterminating animals like the white rhinoceros and the American bison by the late nineteenth century. If we had, therefore, come upon the last *Hydrodamalis* population just a hundred years later, I tell myself, we might have been able to restrain ourselves from wiping it out and even managed to return it, eventually, to the Big Sur, like we've returned the white rhinoceros to the Kruger Park.

The sobering reality of my species' relationship with the Sirenian order is, however, that the population levels of its four living species (three manatees and one dugong) are dropping steeply under our destructive impact.

CHAPTER 7

South America

SOUTH AMERICA HAD, LIKE AUSTRALIA, BEEN AN ISLAND CONTI-
NENT until it joined up with North America some 3.25 million
years ago. It had originally gained its island status by separating
from Antarctica some 30 million years BP. (Australia had broken
away from the other end of Antarctica about 10 million years
earlier.)

After South America went its own way, many of its plants and
animals still had close relatives in Antarctica (which had remained
a temperate, forested continent until at least 20 million years ago).
South America's plants and animals also retained close connections
with those of Australia. As we'll see in Chapter 20, South America
and Australia still share closely related plants, invertebrates, fishes,
amphibians and reptiles. Marsupial mammals still live in South
America as they do, of course, in Australia. (One of the South American marsupials, the "Virginia" opossum, has very successfully
invaded North America.) South America was, therefore, a kind of
"almost-Australia." In addition to its marsupials, however, it also
had placental or "eutherian" mammals (i.e., members of the group
that humans, bats and whales belong to).

Marsupials and eutherians had divided South America between them in an unexpected way—eutherians filled most of the continent's plant-eating niches, while the marsupials were, in the main, flesh-eaters. One of the marsupial flesh-eaters—a member of the so-called "borhyeanid" family—developed a cat-like skull and "saber" teeth that bore an extraordinary similarity to those of the eutherian sabertooth *Smilodon*.

South America's marsupials were not, however, that continent's only flesh-eaters. They competed with several "phororhacoid" bird species, the largest of which, *Titanis*, was a ten-foot-tall, 350-pound monster with a huge head equipped with a fifteen-inch-long hooked beak. *Titanis* was probably able to kill horse-sized prey. "Never before or since," E. O. Wilson tells us,

> have mammals faced anything like the phororhacoids, except during their earliest evolution in the Age of Dinosaurs. In South America Titanis and its relatives must have been serious rivals to the borhyeanids and other carnivorous marsupials. Since anatomists consider birds as a whole to be direct descendants of dinosaurs...the phororhacoids might be called the final echo of the ruling reptiles.

All the marsupial predators and almost all of the phororhacoid bird species were driven into extinction soon after big cats and other North American predators entered South America as a result of the land connection that arose between North and South America. (The entry of North American species into South America, and vice versa, that resulted from this connection is sometimes referred to as the "Great American Biotic Interchange" or "GABI.") A few vertebrae, two cheekbones, and some wing and foot bones found in 2-million-year-old sediments in Northern Florida show that at least one of the phororhacoids—*Titanis* itself, no less—had not only survived the GABI but actually gone on to invade, and establish itself in, the territory of the supposedly "superior" North American predators.

Titanis would become extinct in the early Pleistocene, some 2 million years before humans arrived in the Americas.

* * *

The diversity of South America's eutherian (i.e., non-marsupial) mammals rivaled that of any of the Earth's other continents. The entirely extinct magnorder Meridiungulata, which was comprised of the groups Notoungulata, Litopterna, Astropotheria, Pyrotheria and Xenungulata, contained a great many species that echoed the shapes of animals that were, like horses, hippos and pigs, familiar to us, and a great many others with forms that we would consider unlikely and outlandish. (Elephants and giraffes might, of course, also seem like unlikely creatures to people who weren't familiar with them.)

Almost all the members of this lost Meridiungulate magnorder disappeared in the GABI. A few genera of notoungulates which included the rhinoceros-sized *Toxodon*, and one of the litopterns (the llama-shaped *Machrauchenia*), are known to have survived that interchange, but these survivors all disappeared in the end-Pleistocene extinction. Paleontologists suspect that *Machrauchnia* may have had a tapir-like trunk. The animal's nostrils opened, however, near the top of its head—above its eyes—rather than at the end of that putative "trunk."

We'll probably never get a realistic idea of the appearance of those end-Pleistocene notoungulates and litopterns. "Getting an idea of what those animals may have looked like," Paul Martin told me, "is like trying to imagine what a rhinoceros would have looked like, if they and all their horse, zebra and tapir relatives were known to us only from fossils."

Martin described these late-surviving meridiungulates as being "among the most mysterious members of the fauna of near time, utterly neglected by a public hooked on dinosaurs, and by Pleis-

tocene paleontologists hooked on mammoths and mastodonts."
I see this neglect in myself, and catch myself drifting into a solip-
sistic delusion that animals so poorly known to us couldn't have
been entirely real. But the notoungulates and litopterns, which still
existed at the end of the Pleistocene, formed by millions of years of
evolution, were of course every bit as real as elands or red kangaroos
are. The first human immigrants to South America must, moreover,
have experienced that living reality.

One other magnorder of eutherians, the Xenarthrans, developed
in South America, and this group does contain members that have
survived into the present. Those survivors are classified into the
orders Pilosa ("hairy ones") and Cingulata ("armored ones"). The
Pilosa are presently represented by five tree-sloth species and
by the giant anteater, while the Cingulata are comprised of eight
genera of armadillos, containing twenty-one species (one of which
has very successfully invaded North America).

Fifteen genera of "ground sloths" were living in South America at
the end of the Pleistocene. These sloths had, therefore, fared much
better against competition from North America's megaherbivores
than most of South America's "autochthons" had. They were not
only holding their own in their South American homeland but
had managed, in addition, to mount a successful invasion of North
America, where four genera were living at the end of the Pleistocene. Nine further ground-sloth genera, which had radiated into
eighteen species, lived on islands in the Caribbean, where they
survived until humans reached their islands thousands of years
after the extinction of their continental counterparts.

The biggest of the ground-sloths, members of the genera *Megatherium* and *Eremotherium*, which grew to the size of an African
elephant, could probably reach as high into trees as Columbian
mammoths could. They would have done so by stretching their
long "arms" aloft while standing upright on a "tripod" consisting
of their massive back legs and their powerful tails. All the ground-

sloth species folded their powerful front claws onto their "palms" or "wrists" when they walked, using the knuckles of those "hands" to make contact with the ground. Ground sloths' massive hind feet were not, however, designed to fold in this way—one family, the *Megalonchidae*, walked on the soles of their back feet, while the other families used the sides of those feet.

The seventy-pound giant anteater *Myrmicophagus* is the nearest thing to a ground sloth that survives on our planet. It walks, like all the ground sloths did, on the knuckles of its front feet, and, like the megalonchid ground sloths did, on the soles of its rear feet.

The giant anteater uses the long, stout claws on its front feet to dig for ants. Ground sloths did not, of course, use their claws for this particular purpose, but some of the prairie- and pampas-dwelling species might well have dug for roots to supplement their diet of shrubs and grasses. Giant anteaters also stand upright on their back legs, using their tails to create a "tripod" the way the ground sloths are thought to have done. One of the reasons why they do this is to make their "arms," "hands" and claws available for defense. *Myrmicophagus* is limber and mobile on its hind legs: it can whip around rapidly to reach an attacker who's approaching from behind. Its claws can be used to rip an assailant, or to hold onto it, in order to crush it with its powerful "arms."

I think it's likely that at least some of the ground sloth species may have defended themselves in much the same way that *Myrmicophagus* does. The ground sloths must, at any rate, have been able to fend off predators in an effective way: they survived, after all, in the presence of the sophisticated North American predators that had supplanted most of South America's native meat-eaters. If we bear in mind, in addition, the fact that some of the ground sloth species that lived in the shadow of those formidable predators did not enjoy a size advantage over them, this argument gains even more force. North America's two biggest ground sloths, the elephant-sized *Eremotherium rusconi*, and the rhino-sized *Param-*

ylodon harlani, were much bigger, of course, than their potential predators, but the ox-sized Jefferson's ground sloth probably weighed about the same as the great *Arctodus* bear. The Shasta ground sloth, which seems to have reached a weight of only about 350 pounds, would have been even smaller than many of North America's saber cats and lions.

Although they would have offered no defense against "stand-off" weapons like spears, ground sloths' claws must have been formidable weapons. I used to think that the claws of tree-sloths were relatively passive hooks, which engage branches in the immobile way a cup-hook holds the handle of a mug. I found out how wrong this idea was when, fishing for tarpon in Suriname in South America, I came upon a three-toed sloth swimming across the Coesewijne River. Out of curiosity, I engaged the claws on one of its front legs with the hook of my fishing gaff. To my surprise, the sloth locked a powerful grasp onto that hook and then started climbing, "hand-over-hand," up the gaff's four-foot shaft. That smooth steel shaft was being clamped so firmly between its claws and the "palms" of its "hands" that no amount of jerking on my part could halt the animal's climb. My boatman had to hit the sloth disconcertingly hard with a paddle to make it let go and get on its way.

This experience was also my first acquaintance with the fact that sloths are good swimmers. Those swimming skills might help to explain the presence in North America of ground sloths several million years before that continent became physically attached to South America and the fact that this family established itself on islands in the Caribbean.

* * *

We'll move, now, from the "hairy" or "Pilosan" order of the magnorder Xenarthrans to the "scaley" Cingulate order. This order (which still contains, as we've seen, twenty-one species of armadillo) was represented, in the late Pleistocene, by several different groups referred to (confusingly) as "giant armadillos." The first of these extinct "giant armadillos," *Dasypus bellus*, was very similar to the living nine-banded armadillo of North America. It may have been about the same size as South America's still-living giant armadillo, *Priodontes giganteus*, a rare animal that can weigh as much as 130 pounds.

The members of a small family of armadillo-like "pampatheres" are also sometimes referred to as "giant armadillos," and they were, in fact, real giants: the only North American representative of this family, *Holmesina septentrionalis*, reached about 600 pounds in weight.

The *Glyptodontidae* is perhaps the best known of the extinct armadillo-like families. Some eighteen genera belonging to it inhabited South America at the end of the Pleistocene. One genus, *Glyptotherium*, had moved into North America. Most glyptodont species weighed over 1,000 pounds. Protected by a "shell" or carapace made up of hundreds of bony polygons bound together by collagen, they must, at first glance, have looked like enormous tortoises. Their heads and tails could not, however, retract into those carapaces. The heads were protected, instead, by bony "helmets," and the tails, by rings of armor near their bases and stiff bony sheaths on their ends. In at least one of the South American species, that sheath was tipped by a spiked ball that made it look like the business-end of a medieval mace.

* * *

South America's recently extinct megafauna includes at least two monkeys that were considerably larger than any other primate, living or extinct, which has been found on that continent. *Protopithecus brasiliensis* and *Caipora bambuiorum*, whose fossil remains have been found in Brazil's Bahia province, each approached fifty pounds in weight. Eckhard Heymann of the Deutches Primantenzentrum thinks that these big monkeys may have been at least partially terrestrial.

Surprisingly, perhaps, the extinct megafauna of the New World also included rodents. South America is still home to giant rodents—the capybara *Hydrochoerus*, with its "loxodont" teeth which we spoke about in Chapter 4, commonly exceeds 100 pounds in weight. During the Pleistocene, a larger capybara, *Neochoerus pinkneyi*, which weighed 150 pounds or more, extended into Florida. *Neochoerus* was not, however, the New World's largest rodent. That title was held by the 450-pound, giant beaver, *Castoroides ohioensis*, which lived only in North America.

As we saw in Chapter 3, the spread of the hominin family out of its African homeland across the rest of the planet has been faithfully tracked by the disappearance of giant tortoise species. The New World was no exception in this regard. One or more Galapagos-sized tortoise species (including the so-called *"Hesperotestudo"*) disappeared in the end-Pleistocene extinction-spasm, as did a large relative of the still-existing sliding turtle *Trychemys*. The bones and shells of tortoises and terrapins (freshwater turtles) are, I should add, commonly found in association with Clovis sites in North America.

* * *

When 73 percent of the North American and 80 percent of the South American mammal species weighing over 100 pounds disappeared at the end of the Pleistocene, at least ten genera of birds in the raptor/scavenger class followed them into extinction. Among these birds were an American version of Africa's marabou stork, several large "New World" vultures, the large condor *Breagyps*, the "American secretary-bird" or "walking eagle" *Wetmoregyps*, and several members of the family *Teratorthidea*, whose largest species, the aptly named *Terratornis incredibilis*, had a sixteen-foot wingspan.

Africa's biggest scavenger/raptor birds are presently threatened with the same fate as these American giants—the lappet-faced vulture, with a wingspan of over nine feet, is declining along with the bigger animals whose carcasses they are adapted to scavenge. Smaller African vultures like the griffon, white-backed, hooded and Egyptian species can and do survive in places where big wild animals have been exterminated, but the lappet-face, whose powerful bill was adapted to deal with the skins and tendons of large animals, fades via scarcity into extinction in such conditions.

When a large animal goes extinct, it doesn't go into oblivion alone. Early in the twentieth century *Buphagus africanus*, the yellow-billed oxpecker (called, more aptly, *geelbek renostervoël* or "yellow-billed rhino bird" in Afrikaans) disappeared from the area which is now protected as the Kruger Park. That disappearance followed the disappearance or decline of many of the area's biggest animals. In contrast to the still-plentiful red-billed ox-pecker, the yellow-bill has evolved a heavy-duty beak which may give it an advantage over the red-bill in relation to the larger species of ticks and flies that live on big mammals like buffaloes and rhinos. A North American bird that may have had a similar adaptation followed the megafauna of that continent into extinction at the end of the Pleistocene: the thick-billed cowbird, *Pyelorhamphus molothroides*.

Yellow-billed rhino birds filtered back into Kruger from Zimbabwe sometime after the middle of the twentieth century and have now, probably because of the return and/or increase in the larger herbivores in the area, started building up their numbers again.

The return of the yellow-bill is a welcome development, but the return of the rhino-dependent ticks was, of course, itself a worthwhile re-enrichment of Kruger's biodiversity. The fact that I'm mentioning just one of those returning ticks—*Dermacentor rhinocerinus*—means only the attractive orange markings and interesting shape of that particular parasite often singles it out for human attention. Also restored to Kruger by the reintroduction of rhinos was one of the world's largest flies, *Gyrostigma pavesii*, whose eggs hitchhiked from Zululand to Kruger on or in their returning hosts.

The pioneers of rhino reintroduction weren't always conscious of the host of satellite creatures they were bringing back to Kruger along with the rhinos:

> Black rhino and fly vector R. dutoiti were locally extinct in K.N.P. [Kruger National Park] (and the old Transvaal) by 1936 from HiP [Hluhluwe-iMfolozi Park]. R. dutoiti was re-introduced with the dung of [a single] translocated female black rhino... proliferating throughout southern K.N.P. from Pretoriuskop to Mooiplaas.

Even under today's conditions, when biologists are more careful about identifying the organisms that can hitch a ride when large animals are translocated, it would be difficult or impossible to make a complete inventory of the host of flies, worms, lice, mites, flukes, fungi, protists and prokaryotes that live around, on, and in rhinos. We can only guess, therefore, at the huge number of small species— mainly but not exclusively invertebrate—that must have followed the extinct members of the Eurasian, American and Australian megafaunas into oblivion.

The "secondary" extinctions caused by the disappearance of those big animals would not, moreover, have been restricted to organisms that were as directly dependent on them as, say, *Dermacentor* ticks, *Gyrostigma* flies or large dung beetles. Any of the big animals or birds that disappeared from the "vanished Serengetis" could have been a "keystone" species whose removal could have led to the extinction of organisms not connected to them in any obvious way. In his *Diversity of Life*, Edward Wilson explains, for instance, that when jaguars and pumas disappeared from Barro Colorado Island in Panama, their prey species, raccoon-like animals called "coatis," and large rodents called "agoutis" and "pacas," experienced a huge increase. Because coatis, agouties and pacas feed on large seeds that fall from the rainforest canopy, they quickly reduced the reproductive ability of trees that produce those large seeds:

Other tree species whose seeds are too small to be of interest to the animals benefit by the lessened competition. Their seeds set and their seedlings flourish, and a larger number of the young trees reach full height and reproductive size. Over a period of years the composition of the forest shifts in their favor. It seems inevitable that the animal species specialized to feed on them also prosper, the predators that attack these animals increase, the fungi and bacteria that parasitize the small-seed trees and associated animals spread, the microscopic animals feeding on the fungi and bacteria grow denser, predators of these creatures increase, and so on outward across the food web back and back again as the ecosystem reverberates from the removal of the keystone species. The extinction of the Pleistocene megafauna could, therefore, have had a significant effect on the diversity of biological communities that supported them. Like the elephants and big bovids which still inhabit the jungles of Africa and South Asia, gomphotheres, ground sloths and toxodons probably created trails, wallows and clearings in the Amazon basin, maintaining, possibly, a more varied mosaic of plant communities than we see there today.

The South American rainforest still constitutes, of course, a staggeringly rich storehouse of biodiversity, but this diversity must, to some unknown extent, have been *greater* when the giant herbivores still formed part of its biome.

THE GROWTH OF OUR AWARENESS ABOUT THE ORIGINS OF THE HUMAN FAMILY AND THE EVOLUTION OF ITS COGNITIVE POWERS

In the first half of the twentieth century, anthropologists were still groping for the basic facts of human evolution. Only in the second half of that century did the identities and attributes of the early members of the human family, and the fact that they had evolved in Africa, finally enter the general scientific consciousness. Consensus is still lacking, here in the twenty-first century, on important issues in human evolution: a long-standing disagreement about our family's early ecological impact remains unresolved, and important pages in the related story of its cognitive ascent, unwritten.

CHAPTER 8

Looking for Our Ancestors

Darwin thought it probable that humans had evolved in Africa. His reasoning went like this: humans are ("however much the conclusion may revolt our pride") members of the "catarrhine" or Old World group of primates. The other group, the "platyrhines," live in Central and South America. The ancestors of mankind must, therefore, have lived somewhere in the Old World other than Australia or the Oceanic Islands. "In each great region of the world," Darwin's logic continued,

> ...the living mammals are closely related to the extinct species of the same region. It is therefore probable that Africa was formerly inhabited by extinct apes closely related to the gorilla and the chimpanzee; and as these two species are now man's closest allies, it is somewhat more probable that our early progenitors lived on the African continent than elsewhere.

When the first remains of an extinct kind of human were discovered, however, they turned up in Europe rather than in Africa. Three years before the publication of *The Origin of Species* in 1859 a skull and some arm, leg and hip fragments had been unearthed near Düsseldorf in the Neander "thal." ("Thal," or "tal" in the modern spelling, related to the English "dale," is the German word for "valley." The valley was named for Joachim Neander, a seventeenth-century churchman. "Neander" is the Roman-Greek version of his original name, "Neumann." Like the original, it means "new man.")

Ancient human bones resembling those found in the Neanderthal had in fact been unearthed elsewhere in Europe since 1829. All the remains of extinct humans discovered in Europe in the nineteenth century were, in fact, of the Neanderthal type. The limb bones of this kind of human were shorter and more massive than ours are, and their lower jaws sloped back steeply under their mouths, leaving no trace of the chin which juts out under the mouths of our kind of human. Under foreheads that receded almost as sharply as their chins, massive brow-ridges topped robust, protruding mid-faces, made more prominent by big, broad noses. The occipital or hindmost areas of their skulls stuck out further than ours do. The shape of the Neanderthal braincase was, therefore, different from ours, but its size was the same or a little larger.

Even in the nineteenth century, it was already realized that Neanderthals must have been in existence relatively *recently*. In anticipation of the discovery of the fossilized remains of the much older species whose body-shape and mental capacity must have stood midway between those of apes and humans, a German disciple of Darwin, Ernst Häckel, suggested a scientific name for it: *Pithecanthropus alalus,* "the ape-man who couldn't speak." Unlike Darwin, Häckel thought that human evolution might have been connected with Asia through a hypothetical, sunken connection between Africa/Madagascar and the East Indies that he called "Lemuria." The East Indies were, after all, also inhabited by "human-like" apes: orangs, gibbons and siamangs. Häckel favored the gibbon as our closest relative. Male chimpanzees and gorillas are, he thought, much bigger than their female counterparts (an assumption that is only true of the latter species), while gibbon males and females are, like those of our species, relatively close to each other in size. Gibbons also walk relatively easily on their hind legs compared to other apes (although they usually hold their immensely long arms above their heads while they're doing so to keep them out of the way). Häckel also attached significance to the fact that the gibbon is—like he assumed *Homo sapiens* to be—a monogamous primate.

A talented Dutch anatomist called Eugène Dubois (1858–1940) was one of the many young Europeans of his generation who were fascinated by the new theory of evolution by natural selection. (*The Origin* had been published a year after his birth.) Initially, Dubois couldn't make up his mind whether Darwin or Häckel was right about where human evolution had started. He eventually settled, however, on Asia. He'd learned from Alfred Russel Wallace's biogeographical writings that Java and Sumatra had been joined to the Asian mainland during the glacial periods. He knew, too, that those islands had never been covered by ice the way large parts of Europe had been. He may have been influenced, also, by the convenient fact that the East Indies, of which Java and Sumatra were parts, were Dutch colonial possessions. Against the advice of family and friends, he resigned a promising academic appointment and joined the Dutch army as a surgeon to enable him to travel, as he did in 1887, to the Indies to look for Häckel's *Pithecanthropus*.

Dubois's first excavations, conducted in Sumatra, were undertaken whenever he could take time off from his medical duties. He was so successful, however, at unearthing the remains of extinct animals that he was eventually able to persuade the Dutch government to turn his paleontological investigations into a formal part of his duties and grant him funds to expand them.

Five years after his arrival in the Indies, excavating a promontory jutting into the Solo River near the Javan village of Trinil, Dubois's workers unearthed the skullcap of a large primate. This skullcap, which included the "supra-orbital margins" (i.e. the top edges of the eyeholes), was complete enough to allow a reasonable estimate of its original capacity: about 850 cc. This was a significant number because it lies about halfway between the average cranial capacity of the bigger anthropoid apes (~450 cc) and that of humans (~1,350 cc). That must have seemed like an indication that the skullcap could be that of the half-human, half-ape species he'd come to the Indies to find.

Dubois was initially inclined, however, to think that the skull had been that of an ape. The fact that its capacity was larger than any of the contemporary ape species might, he feared, simply have reflected the fact that it was merely that of an exceptionally large kind of ape. A year later, however, and some fifteen meters from where the skullcap had been found, Dubois's workers found a thigh bone—a femur—that was indistinguishable from that of a modern human. The Trinil site had also yielded a molar tooth which was completely human in appearance. Because the femur and the molar were found in the same layer of gravel that had contained the skull-fragment, Dubois concluded (correctly, as subsequent finds would show) that all three were remains of the same species. The human characteristics of the latter two finds persuaded him that the species in question wasn't the "pure ape" that he'd originally feared it might be, but that it was, in fact, the ape-man he'd come to the East Indies to find.

Because its very human-like femur suggested strongly that his "ape-man" had walked upright, Dubois named his find *Pithecanthropus erectus* (rather than *P. alalus*). It's ironic that Dubois initially feared that the skull he'd discovered was too ape-like to be the *Pithecanthropus* he was looking for because *erectus* turned out, in the end, to be too human to be an "ape-man." The size of its brain was, as we've seen, only about halfway between that of our brains and those of the great apes, but that was the only major difference between it and *Homo sapiens*. Subsequent discoveries would show beyond doubt what the Trinil femur and molar tooth already hinted at: that the rest of their owner's anatomy was overwhelmingly human-like. The species discovered by Dubois is now referred to, therefore, as *"Homo" erectus* rather than *Pithecanthropus erectus*. The rules of scientific nomenclature tell us that the "erectus" part of its name is here to stay, but it, too, has been shown to be inappropriate: later finds have made it clear that *erectus* was by no means the first member of the human family to walk on two legs.

In the light of the many discoveries made since Dubois's death in 1940, it's clear that many of the assumptions he made about his 1891 discovery were wrong, but equally clear that his pioneering quest to find the bones of a human species older than *Homo neanderthalensis* had succeeded. After Dubois returned to Europe in 1895, he exhibited his finds at a series of lectures and discussions.

Enter Rudolf Virchow, a famous German pathologist who'd become widely accepted as the leader of, and spokesperson for, the German scientific community. (Unappreciative members of that community referred to him as "the Nabob" or "the Pasha.") Virchow, an anti-evolutionist, had previously dismissed the Neanderthal finds as the bones of modern humans that had been distorted by disease. Now he rejected the idea that Dubois had found a human-like being as "the purest fantasy imaginable." The remains Dubois had dug up in Java were, Virchow explained, those of a giant gibbon.

Seen in a wider context, Dubois's battle with Virchow was just another action in the war that was being waged at the time between scientists who accepted Darwin's views and those, like Virchow, who didn't. Dubois was not inclined to adopt a detached perspective on Virchow's dismissal of his extraordinary discovery. A proud man, sensitive to the point of paranoia, Dubois was short on both tolerance and what we'd now call people skills.

Even those who agreed with his proposition that *Pithecanthropus* was not an ape could pose a threat to Dubois's peace of mind. A Strasbourg anatomist named Gustav Schwalbe traveled to Leiden to examine the Trinil find and became convinced, as a result, that *Pithecanthropus* was the beginning of an evolutionary sequence that had run through the Neanderthals to culminate in *Homo sapiens*.

The publication of this theory advanced both Dubois's cause and the new science of paleoanthropology, but it also provided a hefty boost for Schwalbe's own professional reputation. In his *Eugène Dubois and the Ape Man from Java*, Bert Theunissen reasons that Dubois resolved, as a result of that boost, not to allow anyone else to examine the fossils until he himself had published a full description and analysis of them. Schwalbe had come close to stealing his thunder, and Dubois did not, Theunissen surmises, want to take the risk of that happening again. This, Theunissen thinks, is the reason why Dubois would not permit anyone else to examine his *Pithecanthropus* fossils after 1900. Dubois's refusal was to remain in effect for nearly three decades.

Remains similar to those of *Pithecanthropus erectus* started showing up in the 1920s and '30s. The first of them came from China. On the basis of a series of finds from Zhoukoudian near Beijing from 1927 to 1933, the Canadian anatomist Davidson Black described a new species called *Sinanthropus pekinensis*. *Sinanthropus* was, in Black's view, very similar to Dubois's *Pithecanthropus*. Franz Weidenreich, who succeeded Black after the latter's premature death in 1933, agreed with this view. Both *Sinanthropus* and *Pithecanthropus* belonged, Weidenreich thought, in the Hominidae or human family, rather than the half-ape, half-human "Pithecanthropoidae" family that Häckel had suggested for the latter.

Dubois reacted to this idea with indignation and tried to refute it in a 1936 article titled "On the Gibbon-like Appearance of *Pithecanthropus erectus*." Nearly forty years before, Dubois had to emphasize the human-like characteristics of *Pithecanthropus* to rebut Virchow's claim that it was "merely" a large form of gibbon. Now he needed to emphasize its supposedly gibbon-like qualities to counter Weidenreich's assertion that it was "merely" a human.

Then new finds were made in Java. In 1936 at Modjokerto, and in 1937 at Sangiran, Ralph von Königswald discovered two more skulls that also seemed similar to Dubois's *Pithecanthropus*. Originally thought to be about a million years old, these are now generally regarded as dating from around 1.49 million years BP. Von Königswald described a species he called *Pithecanthropus modjokertensis* on the basis of the first of these finds, provoking a vehement objection from Dubois. The Trinil primate—his primate—was, Dubois insisted, the only member of the Pithecanthropoidae to have been discovered up to that time. Von Königswald relented and renamed the species *Homo modjokertensis* (but continued, we're told, to refer to it as *Pithecanthropus* in private). The earhole and part of the jaw of the Sangiran specimen—areas that had not been preserved in Dubois's find—demonstrated beyond question that its owner had been a very human-like being. Von Königswald and Weidenreich now made a detailed comparison between the Javan and Chinese material and came to the conclusion that all the Asian finds, including Dubois's so-called *Pithecanthropus*, were "related to each other in the same way as two different races of present mankind."

By the second half of the twentieth century, all the Asian finds including Dubois's were being referred to as *Homo erectus*.

Dubois himself agreed that Davidson's *Sinanthropus* was a member of the Hominidae; he thought it was a kind of Neanderthal. He was also comfortable with (and indeed, as we've seen, insistent on) the idea that von Königswald's Javan finds were members of the human family. To the day of his death in 1940, however, at the age of eighty-two, he clung to the belief that his find was not a member of that family, but of the ape-man family "Pithecanthropoidae."

* * *

Among the first scientists to support Dubois's 1895 claim that his Javan primate wasn't "just an ape," were a Scot called Arthur Keith and an Australian called Grafton Elliot Smith. Both these men were young near the turn of the twentieth century, and both, as one might suspect from their receptivity to Dubois's claim, were evolutionists. Keith didn't only accept the fact that *Pithecanthropus* had human-like attributes—he also accepted the idea that it had been completely adapted to upright walking.

The fact that it had a comparatively small brain persuaded Keith, however, that *Pithecanthropus* must have occupied a side-branch of the human family. Like many of the anatomist-anthropologists of their time, Keith and Elliot Smith thought that the ancestor of *Homo sapiens* must have developed a big, intelligent brain before it had become capable of walking upright. "It was not," as Elliot Smith put it, "the adoption of the erect attitude that made Man from an Ape, but the gradual perfecting of the brain, and the slow upbuilding of the mental structure, of which erectness of carriage is one of the incidental characteristics." Keith was in agreement with this view:

> He [Elliot Smith]... rightly foresaw that before the anthropoid [i.e., ape-like] characters would disappear from the body of primal man, the brain, the master organ of the human body, must first have come into its human estate. Under its dominion, the parts of the body such as the mouth and the hands, the particular servants of the brain, became adapted to higher uses.

In addition to sharing the then-current "big brains came first" view of human evolution, Arthur Keith also believed in what he called "man's antiquity"—the idea that *Homo sapiens* had evolved its "true human" or "modern" form much further back in time than was generally believed. He believed, therefore, that a big-brained being, very similar to *Homo sapiens*, had already been walking upright before either the Neanderthals or Dubois's small-brained *Pithecanthropus* came into existence.

He supported this notion by suggesting that several of the modern-looking skulls that had been discovered in England itself—most notably the Galley Hill skull that had been found in 1888—were extremely ancient. But Galley Hill and the other supposedly "ancient" British remains did not, in fact, possess any features that would support the idea of "the vast antiquity of true humans." Apart from a somewhat greater-than-average thickness in the Galley Hill skull, the "primitive" features that Keith saw in it simply weren't visible to his fellow anatomists. Galley Hill and all the other remains for which Keith claimed "great antiquity" are now universally viewed as recent material.

* * *

In 1907 a human fossil that really *was* extremely old turned up in Germany. After unearthing mammalian remains that included many extinct mammals in the district of Mauer over a period of some twenty years, Otto Schötensack, a professor of geology at nearby Heidelberg University, found a big, robust lower jaw that lacked the kind of protruding chin that *Homo sapiens* possesses. The big jaw was beautifully preserved, and structures like the symphyses or joins between its constituent parts confirmed the already-clear message conveyed by its teeth: its owner had been human. The level at which the jaw was found, and the kinds of animals that were found with it, made it clear, though, that the Mauer human had lived long before the Neanderthal form emerged. (The Mauer jaw is presently thought to be about 600,000 years old, while the Neanderthal form only appeared about 300,000 years ago.)

Schötensack named a new human species on the basis of the jaw: *Homo heidelbergensis*. No skull parts or other human bones were found in association with the jaw, so its finders could not tell whether *heidelbergensis* had either a "modern" skull capacity or a "modern" ability to walk upright.

Then, in 1912, an even more startling find was made—this time in England. Human skull fragments were found together with an ape-like jaw fragment in a gravel pit on a farm near Piltdown Common in the county of Sussex. The ape-like jaw-fragment bore two seemingly human-like teeth. Most people concluded from this, and from the fact that the skull fragments and jaw had reportedly been found so near to each other, that the skull and the jaw must have belonged to the same species—and, in fact, to the same individual. The skull fragments were clearly those of a human species, but that human species must, judging by its ape-like jaw, be an older one than *heidelbergensis* with its obviously human-like jaw. This reasoning emboldened Arthur Smith Woodward, keeper of geology at the British Museum of Natural History, to create a new genus with the presumptuous name *Eoanthropus*—"Dawn Man"—for the Piltdown find.

This impression of great age was reinforced by the remains of extinct animals (mainly tooth fragments) that were associated with the Piltdown find: two fragments of molar teeth represented what was apparently a Pliocene mastodont, and the third, a Pliocene stegodon. Tooth fragments representing hippopotamus and beaver species that could also have lived in the Pliocene were found in the same deposit. The Piltdown pit where these finds had been made also yielded stone and bone implements from a seemingly early stage of human technology.

Woodward had been present when the jaw had been dug out of the ground by his friend Charles Dawson, a solicitor who was an amateur archaeologist and paleontologist. Early in 1912, Dawson had written to him that workmen had—some years previously—unearthed, and broken, a human skull in a gravel pit on a farm near Piltdown Common in the county of Sussex. A short while after he'd sent this report to Woodward, Dawson brought a fragment or fragments of the skull to London. The bone was extraordinarily thick. "How's that for Heidelberg!" he is reported to have called out as he handed the material to Woodward.

Woodward took the bait, and, over most of the weekends in the summer of 1912, he traveled to Sussex to search, together with Dawson, through material dug from that pit. On a few occasions, the pair was assisted by a thirty-one-year-old French priest by the name of Teilhard de Chardin who both of them had known previously. Dawson found most of the rest of the human and animal remains that came out of the Piltdown gravel pit during that summer, including the ape-like jaw fragment, but Teilhard and Woodward also made a few finds.

Placing *Eoanthropus* as far back as the Pliocene would not only confirm that it had lived before *Homo heidelbergensis*—it would suggest, in fact, that *Eoanthropus*'s age must rival or exceed that of Dubois's *Pithecanthropus*. Notwithstanding this, however, the Piltdown skull had clearly housed a brain that was about the same size as that of a modern human. If a big brain had been the first distinctively human characteristic to develop, the *Eoanthropus* must surely have a better claim to human ancestry than the small-brained *Pithecanthropus*.

A number of anthropologists doubted, however, that so "modern" a human skull could have been combined in the same species with such an ape-like jaw. These doubters—the so-called "dualists"—felt that the Piltdown skull and jaw must represent two different species: a human and an ape. An American paleontologist even created a new species for the presumed ape: *Pan vetus*—"ancient chimpanzee." The feeling among these dualists was that the skull was probably an "intrusive" object—that is, one that might have been buried in the ancient deposit containing the ape-like jaw at some relatively recent time. But if that were so, then why were the walls of its skull as thick as, or thicker than, those of a Neanderthal skull?

As the controversy over Piltdown burned bright, the careers of those associated with the find were brilliantly illuminated. Woodward had enjoyed his share of professional honors before Piltdown—he'd been made a Fellow of the Royal Society on the basis of his work in marine paleontology as early as 1904—but his close association with Dawson's find turned him into a scientific celebrity. Taking part in that find was, as he described it to a newspaper reporter in 1924, "the most important thing that ever happened in my life."

Arthur Keith, who had played a prominent part in the evaluation and reconstruction of the skull, received an even bigger boost from its "discovery." Keith's reconstruction of *Eoanthropus* became the centerpiece of the anthropological exhibit at the museum of which he was the conservator. People flocked to Keith's museum lectures in record numbers. Piltdown wasn't only the centerpiece of those lectures and exhibits—it became the centerpiece of Keith's career.

Before Piltdown, that career had been in the doldrums. In 1900 Keith had written a book titled *Man and Ape—a Statement of the Evidence of Their Common Origin as It Stands Today*. After the manuscript was completed his publisher, John Murray, had second thoughts about its salability. Murray praised the book's scientific merit but refused, despite the fact that he'd signed a contract with Keith, to go through with its publication.

"Murray's rejection of the book," Keith would later write, "was my bitterest disappointment in my struggle for place and reputation among my fellow anatomists." During the next eight years, Keith worked as an anatomy demonstrator, publishing some work on the human heart. Finally, in 1908, he was appointed as the Conservator of the Hunterian. This was a big step up for Keith, but the new job was still, in anthropologist-historian Pat Shipman's words, "a relatively obscure post that he found unsatisfactory."

Keith would have to swallow more disappointments before his career would take off. In 1911, when he was already forty-five years old, the Royal Society gave him what he described as a "slap in the face" by rejecting his application for membership. (Elliot Smith had, by contrast, been given a Fellowship of the Royal Society—the coveted "FRS"—in 1907, at the age of thirty-six.) Keith reapplied in 1912, a few months before the Piltdown discovery was announced and was again rejected.

It was only in 1913—after his association with Piltdown had thrust him into the public eye—that Keith was finally admitted to the Royal

Society. The *Man and Ape* book that John Murray had rejected now metamorphosed into a new work titled *The Antiquity of Man*. Published in 1915, *Antiquity* was dominated by discussion of Piltdown. An image of a skull titled "Piltdown Fragments" was embossed in gold on its front cover. In 1921, Keith was given a knighthood. "He has gone up," Oxford Professor of Geology, W. J. Sollas, wrote in 1924, "like a rocket and will come down like the stick."

* * *

Piltdown was, of course, exposed as a fraud in 1953. Charles Dawson the solicitor (who practiced at Uckfield near the Piltdown site in Sussex at the time of the "discovery") was clearly the perpetrator of that fraud. John Walsh's 1996 book *Unraveling Piltdown* gives a carefully substantiated and devastating account of the string of faked archaeological and paleontological "discoveries" that Dawson concocted before he engineered the Piltdown "discoveries." Dawson added old-fashioned business fraud and plagiarism to those scientific frauds. Walsh's book gives Dawson's responsibility for the Piltdown fraud what Richard Bernstein called, in a 1997 *New York Times* review, "the solidity of a Euclidean proof."

The skull of Dawson's supposed "dawn man" was, in reality, that of a recently living human. Its greatly thickened condition was the result of a disease called rachitis or rickets. The jaw fragment was that of a young female orangutan that had been alive even more recently than the owner of the skull. The characteristic high cusps of the two teeth it contained had been filed down to give them a human appearance. At the start of his investigation, Joseph Weiner, a member of the three-person team that proved that Piltdown was a fake, suspected only that Piltdown's ape-like jaw was a forgery or "plant." He was astonished, therefore, to discover that *none* of the thirty-seven pieces of bone and stone "found" at Piltdown were genuine. They had all been brought from elsewhere, and all chemically stained reddish-brown to match the color of the gravel in the pit in which they were "found." Ten other faked objects were, in addition, "found" at two other sites near Piltdown.

The second of these two subsequent "discoveries" consisted of human skull fragments, which Dawson reported finding in 1913 at Barcombe Mills (some four miles away from Barkham Manor, the original Piltdown site). Nobody challenged the genuineness of this "find," but Arthur Smith Woodward didn't announce it as another specimen of *Eoanthropus*.

"It is *not* a *thick* skull," Dawson had written to Smith Woodward, "but it may be a descendant of *Eoanthropus*."

The third human "find" was reported by Dawson at the beginning of 1915. This one consisted of two cranial fragments, an altered molar from the lower jaw of an ape (which probably came from the same orang jaw "found" at the first site), and a fragment of a rhinoceros molar. This time Dawson was able to claim confidently that he'd found another *Eoanthropus*: "...the general thickness," he wrote, announcing the "discovery" of one of the skull fragments to Smith Woodward, "seems to me to correspond with the right parietal of *Eoanthropus*..."

Dawson told Woodward that this latest group of bones was found "on the Sheffield Park Estate." He would not, however, tell Woodward the exact place where he found them. Two full years were to go by before Woodward would announce the Sheffield Park "find." Had a kind of standoff developed between the two men, with Woodward saying he wouldn't announce the find until Dawson told him exactly where he'd found it, and Dawson threatening to make it public via some other member of the scientific community if Woodward didn't stop delaying? We know that Dawson did tell Ray Lankester (an Oxford professor of anatomy, later attached to the Natural History Museum) about the new "find," because the latter inserted a footnote in his 1915 *Diversions of a Naturalist*, which stated that Dawson had discovered "a second skull of the same character as the first."

Woodward's wife later vividly recalled this uneasy and troubled interlude, explaining that Dawson "would not give details of the exact spot" and that her husband on his own "spent much time searching for it" fruitlessly. Toward the end of 1915 Dawson became ill with anemia. In August of 1916, he died at the age of fifty-two. Dawson's widow passed the Sheffield Park bones and teeth to Woodward. Woodward questioned her, and other associates of Dawson, in vain to get a better idea of where Dawson had found them, delaying the announcement of "Piltdown Two" for another six months after Dawson's death. He must have been agonizing about whether he should make that announcement at all. If he decided not to do so, he would have to explain why he wasn't doing so—Dawson had, after all, already leaked the news of the "discovery" before his death. If he made the announcement, on the other hand, and qualified it with an honest admission that Dawson had refused to tell him exactly where he'd found the material, he would not only have raised a big question mark in relation to "Piltdown Two," but would also have aggravated the doubts that many people were already feeling about Piltdown One.

In the end, the benefits of covering up the truth proved to be too tempting for Woodward. At a meeting of the Geological Society on February 28, 1917, he reported that a second set of *Eoanthropus* remains had been found on a "large field, about 2 miles from the Piltdown pit." He implied that he knew where that field was by adding that he and Dawson had already examined it "several times without success during the spring and autumn of 1914." "Not then, not ever afterward," Walsh tells us, "did Woodward further identify or even refer to the 'field' he specified so prominently in his official report."

The announcement that a second thick-boned skull had been found at a separate location, together with a lower molar tooth very similar to the teeth contained in the ape-like jaw of "Piltdown One," silenced the critics who were maintaining that the skull and jaw of Piltdown One could not represent a single species. "Gradually," Walsh tells us, over the next several months and years, accelerating with the war's end in November 1918, many experts who had vehemently opposed the combination of jaw and cranium now simply gave way, confessing themselves wrong. The conversion, while by no means complete, was dramatic and sweeping, affecting some of the leading names in the field worldwide...

* * *

The Piltdown fraud was emphatically not, as the late Stephen Gould argued, a harmless joke that developed into an embarrassment for its perpetrators. Much more time, effort and care were devoted to it than are spent on what we usually think of as pranks. As we've seen, the fraud didn't just consist of planting a single ape jaw fragment with pieces of a human skull—it involved the appropriation and modification of a much larger number of specimens. Some forty-seven pieces of bone and stone were assembled, broken, shaped and/or stained in the course of perpetrating it. These pieces were then "found," or planted for others to find, at Piltdown or reported by Dawson as having been found elsewhere, over a period of three years. The intense and sustained nature of this effort makes no sense unless we accept that it was directed at the attainment of some non-frivolous, non-trivial goal.

That goal could only have been the achievement of scientific recognition (together, of course, with the social and financial benefits that such recognition would entail). None of Dawson's previous scientific frauds and fakeries were "pranks"—all of them were squarely and rationally aimed at obtaining scientific credit. Dawson did, in fact, reap a considerable degree of acclamation for his "discovery" of *Eoanthropus dawsoni,* but not as much as he'd hoped for. He was turned down for the FRS despite the fact that his name was put up for consideration each year between 1913 and his death in 1916.

Nearly everyone who was directly involved with Piltdown (and several others on the periphery) have been suspected of conspiring with Dawson to commit the fraud. During the 1990s Arthur Keith was in fashion as a suspect, with two scholars, Phillip Tobias and Frank Spencer, building detailed cases against him.

Keith was as capable of covering up truths that may have cast doubts on Piltdown's authenticity as Woodward was. Although the Piltdown skull did not, for instance, display the gross deformation that is sometimes brought about by rickets, it did bear less prominent indications of having been affected by that disease. The first person to spot those indications seems to have been S. G. Shattock, who was in charge of the largest section of the Hunterian Museum, which was concerned with pathological lesions of the human body. Shattock examined the Piltdown skull fragments soon after they were "found," and concluded, in a 1914 Hunterian publication that

> ... certain details of the Piltdown calvaria [top of the skull]... suggest the possibility of a pathological process having underlain the thickened condition.

In an attempt to establish the specific nature of the "pathological process" that may have thickened the skull, Shattock excluded the possibility of acromegaly and six other potentially bone-deforming conditions. He stated specifically, though, that he could not exclude "a past rachitis that has been followed by a reconstruction of the bone." A year later, however, commenting on what he called the "surprising" thickness of the Piltdown skull, Arthur Keith made the following assertion in his *Antiquity of Man*:

> The bone is naturally formed; there can be no question of disease. My colleague Mr. Shattock definitely settled this point.

That statement is on a par with Woodward's lie about the "large field" in which Dawson was supposed to have found Piltdown Two. Neither of these false statements constitutes proof of an active conspiracy with Dawson, but both were calculated to hide such manifestly suspicious aspects of the Piltdown "find," that one feels that the two men who told them were guilty of something almost as serious.

* * *

Paleoanthropology can be a bruising business. We've seen how it led to disillusionment and isolation for Eugène Dubois. It can, however, also produce very big rewards. In a 1954 letter to Ashley Montague, the eighty-eight-year-old Arthur Keith talked about those rewards, prefacing his remarks with a description of the "finder" of the Piltdown skull. "So compelling," he wrote, "was the honesty of Dawson's manner of speech that not a single soul of us, doubted his word, yet," Keith continued,

> I have now no doubt that he was the author of all the fraud. And you will ask—what could have been his motive? If you knew the wonderful fame won by Shoetensack in 1907 by the discovery of the Heidelberg jaw—you would realise the fame waiting for the discoverer of a skull of that early date.

As one of the chief beneficiaries of Dawson's "discovery," Keith had, as we've seen, plenty of personal experience with the "wonderful fame" important finds in the field of human paleontology could generate.

In the 1920s, many if not most anthropologists regarded "Piltdown man" as a direct ancestor of our species, and the oldest such ancestor, moreover, to have been found to date. The faked human species fit in perfectly with widely accepted beliefs of its time. It supported both the "big brains came first" view and a "Europeans came first" belief. It appeared at a time when European civilization in general, and the British Empire in particular, could still look like the crowning achievements of human evolution. It made sense to many that the evolution of our species must have begun in the place where those achievements were realized. How else could one explain why Western Europe—and England in particular—seemed to have enjoyed a head start in the race for political, technological and cultural ascendancy?

All these comfortable assumptions were to be challenged in 1924 when a young Australian called Raymond Dart discovered a human-like being that was clearly more ape-like than Dubois's *Pithecanthropus* in the most "uncivilized" of all the continents: Africa.

CHAPTER 9

Ex Africa Semper Aliquid Novi

Rɪɢʜᴛ ꜰʀᴏᴍ ᴛʜᴇ sᴛᴀʀᴛ ᴏꜰ ʜɪs ᴄᴀʀᴇᴇʀ, ᴛʜᴇ Uɴɪᴠᴇʀsᴇ sᴇᴇᴍᴇᴅ to have been nudging Raymond Arthur Dart toward a study of the anatomy of the human brain. That's the way you might look at it, anyway, if you were inclined to mystical thinking. The nudging got under way in 1914 when Dart, a twenty-year-old medical student in Australia, attended a public lecture titled "The Evolution of the Brain." The speaker was Grafton Elliot Smith who was then one of the world's leading authorities on the comparative anatomy of the brain.

"I fell under his spell that night," Dart wrote later, "and prayed that at some time I would be allowed to work under him." While he was finishing his medical training in Sydney during the First World War, Dart was appointed as an assistant to James T. Wilson, a professor in anatomy who also had a special interest in the development of the brain. "Like Darwin," Dart tells us, "Wilson was fascinated by vestigial structures and the light they threw on the grand evolutionary story of the brain." From 1915 to his graduation in 1917, Dart enjoyed a "treasured and intimate" relationship with Wilson. "His influence on me was so great," Dart wrote at the age of sixty-five, "that even today I often find myself guided by the standards which he implanted in my young mind."

Toward the end of the First World War, Dart served as an army doctor in France. Then, through his connection with Wilson back in Australia, he managed to land a job in England, working for Grafton Elliot Smith himself. "Working under Elliot Smith," Dart wrote years later,

> was my student dream come true. Not only was he a genius in his own field but one of the most pleasant human beings I have ever worked for or with. Tall, ruddy-complexioned and distinguished, with immaculate white hair, he was the complete antithesis of the woolly-minded, innocent genius of fiction. Elliot Smith was with all his brilliance, in every sense, a man of the world, a great raconteur and popular with his colleagues and assistants who could usually rely on him to attend and enliven their daily tea parties.

While Dart was working for him, Elliot Smith appointed Nikolai Kulchitsky, a former Minister of Education under the czar, as a laboratory assistant in an emergency measure to help the latter keep body and soul together. Having this sixty-four-year-old refugee from the Bolshevik revolution working for him in such a humble capacity could have been an awkward situation for Dart. Kulchitsky was a world-famous authority on the histology or microscopic structure of the nervous system. The interaction between the two men turned out, however, to be pleasant—and fruitful for Dart. Kulchitsky, as ignorant of English as Dart was of Russian, used bits of French and German to make an important contribution toward the latter's knowledge of the histology of the brain and other parts of the nervous system.

The year 1922 saw Elliot Smith working on the reconstruction of the Piltdown "find." The fact that his chief's attention was now focused on anthropology gave fresh impetus to Dart's enthusiasm for comparative brain anatomy, and he spent all his free time working his way through the comparative collections in the museum of the Royal College of Surgeons (of which Arthur Keith was, as we've seen, the Conservator). That collection included "endocranial casts," or "brain casts," of humans and of apes. Dart himself had learned to make brain casts as part of the work he was doing for Elliot Smith— it's done by pouring a liquid, usually latex, into an empty skull and then removing it when it has solidified into a mass that accurately reflects the shape of the skull's original contents.

In the next year, 1923, the twenty-nine-year-old Dart left England, with his American wife, Dora, to take up a professorship in anatomy at the newly created University of the Witwatersrand in the mining town of Johannesburg in South Africa. Dart was not going off to this distant outpost of the empire willingly. Elliot Smith had, in fact, pushed him out of the nest—in addition to recommending him for the South African job, it seems that he'd put Dart on notice that his English job would no longer be available. South Africa was, at first, a terrible disappointment. The medical school's buildings were dreary and inadequate, and equipment was basic or nonexistent. Incredibly, there wasn't even a "general" library at the university, let alone a medical or anatomical one.

Dart made the best of things. He created an embryonic library, and, thinking perhaps, of the wonderful museum of the Royal College of Surgeons in which he'd spent so much of his free time, he offered five pounds—a sum equal to several hundred present-day US dollars—to the student who found the most interesting fossil during each annual vacation.

One of his students, Josephine Salmons, told him about a fossil skull she'd seen at a family friend's house. It had been blasted out of the Buxton lime-mine about 200 miles west of Johannesburg near a place called Taung. (Pronounced ta-OONG, the name means "place of the lion" in Setswana.) Salmons thought the skull might be that of a baboon. Not wanting to dampen her enthusiasm, Dart told her as

gently as he could that she was probably wrong. "Other than Rhodesian Man and Boskop Man," he explained, "no single fossil of any of
the primates... has ever been reported south of the Fayum deposit in
Egypt."

But it was Dart who was wrong. Hans Reck, a German vulcanologist, paleontologist and adventurer, had found baboon fossils at
Olduvai Gorge just before the First World War, and baboon fossils
had already been reported from South Africa itself. Like many
scientists of that time, Dart still shared the prevailing view that
the "primitive" or "backward" continent of Africa could not have
been the ancestral home of the most advanced of the mammalian
orders, the primates. (The "Rhodesian Man" that Dart mentioned
to his student was considered to have been a relative newcomer to
Africa, one whose ancestors had wandered down from Europe near
the tail-end of the Neanderthal period.)

* * *

The skull that Salmons brought to the university the next day was,
at any rate, that of a baboon. Apparently Africa wasn't as devoid
of primate fossils as Dart had assumed it to be. "Within minutes,"
Dart tells us, "I was careering down the hill in my Model-T Ford to
discuss the skull and Taung with a friend and colleague, Dr. R. B.
Young, a veteran Scottish geologist." Young was, at the time, doing
some work for the owners of the Buxton mine. At Dart's request, he
spoke to the manager about preserving fossils found during mining
operations. That, Young was told, was already being done. One of
the miners, a Mr. de Bruyn, was an amateur fossil collector, and he'd
just brought several fossils and fossil-containing fragments of limestone to the manager's office. These, the manager promised, would
be packed up and sent to Dart.

The fossils arrived at Dart's house in two boxes just as he and his
wife were making final preparations for the wedding, at their
home, of a close friend. As Dart told the story years later, his wife

and the groom became anxious when he tore off the formal collar he'd affixed for his role as best man and started opening the boxes. However high Dart's expectations of the Taung fossils may have been, they were exceeded by what he found in the second box.

Lying on top of the rubble was something he'd become familiar with in England—an endocranial cast. This particular one was a "natural endocast"—that is, one that had not been created by a human anatomist but by the natural seepage of a cement-like mixture of lime and sand into a skull. The skull in question had (Dart realized) been that of a primate. That primate had, moreover, been an extraordinary one. Its brain had been fully three times the size of a baboon's brain. The brain was, in fact, somewhat bigger than a chimpanzee brain and a bit smaller than that of a gorilla—about a third of the human size. The brain cast Dart was examining was, therefore, at least that of an ape (rather than a monkey or a baboon) and was, as such, already a very significant find.

But that cast *didn't* just represent the brain of an ape—not an ape, at any rate, of the gorilla-chimpanzee-orangutan kind. Dart tells us that he realized right away, in the few minutes he was stealing from the wedding preparations, that it had been formed in the skull of a creature that was more "human-like" than the living apes. To understand how he could have come to that conclusion—and come to it so *quickly*—we have to get an idea of how informative endocranial casts can be. The inside of a mammal's skull isn't smooth like, say, the inside of a porridge bowl. It's a porridge bowl, instead, whose inside surface has taken on the shape of the lumps in the "porridge" it contained. Plainly visible, therefore, on the endocast Dart was looking at, was a "startling image of the convolutions and furrows of the brain and the blood vessels of the skull."

Among the furrows that run across the sides of the brains of apes is one called the "parallel sulcus." Not far behind it, is the so-called "lunate sulcus." ("Sulcus" is Latin for "ditch.") In humans, these two "sulci" are much further removed from each other. Expansion of the temporal and parietal lobes of our brains has pushed the lunate sulcus well to the rear of the position it occupies in apes, and downward to some extent. In the process of doing this, the parietal cortex has almost buried the lunate sulcus. Dart's old chief, Elliot Smith, had, in fact, been the first person to establish the existence of the lunate sulcus in the human brain. It was Elliot Smith's investigations that had shown that, in the rare cases where the lunate sulcus *does* show up on the surface of the human brain, it will be seen *well to the rear* of the parallel sulcus.

Dart was able to satisfy himself at first glance that the lunate and parallel sulci on the Taung endocast were, in his words, "separated by a distance *three times as great* as in any existing endocast of a living ape's skull, whether chimpanzee or gorilla."

After "ransacking feverishly" through the boxes, Dart found a large rock with a hole in it that fit the endocast perfectly. A bit of skull and lower jaw were faintly visible in the rock, and Dart felt confident that it contained at least part of the skull or face that went with the endocast. That face would only emerge weeks later after much patient chipping and scratching, but Dart had already satisfied himself on the basis of brain anatomy that he'd found the fossil of a human-like ape. The phrase "human-like ape" can, however, be misleading in this context. Chimps, gorillas, orangs, gibbons and siamangs are already classified as "anthropoid"—that is, "human-like" apes. What Dart had recognized was something *more* human-like than the so-called "anthropoid" apes.

The fact that the Taung endocast possessed human-like characteristics was, moreover, an *obvious* one for Dart. "I knew at a glance," he wrote later, "that what lay in my hands was no ordinary anthropoidal [i.e., ape-like] brain." In making this determination, he seems to have relied as much on the general shape of the brain as on the rearward displacement of the lunate sulcus. It was, he wrote, "a big, bulging brain and, most importantly, the forebrain was so big and had grown so far backward, that it completely covered the hindbrain."

When a contemporary anatomist uses the term "forebrain," he or she is usually talking about the entire cerebrum, as well as the thalamus, the hypothalamus and the limbic system. When Dart tells us, however, that the "forebrain" of the Taung species had expanded and grown backward, he's talking only about the anterior lobes of the cerebral cortex. In both apes and humans (as well as the other mammals), the cerebral cortex has four lobes in each hemisphere of the brain: the frontal, parietal, temporal and occipital. In humans, the frontal, parietal and temporal lobes have expanded to crowd the occipital lobe into a smaller area at the back of the brain. This is why the lunate sulcus, situated near the front of the occipital lobe, has been pushed so far to the rear in our species.

The Taung endocast was also laterally compressed—the skull that enclosed it must, in other words, had to have been a fairly *narrow* one. While ape skulls have a relatively low dome that gives them a bun-like shape (viewed from the front or the back) human skulls viewed from those positions have steep, almost vertical sides that give them a "loaf of bread" shape.

Dart was, therefore, reacting to several cues when he realized "at a glance" that the Taung endocast had human-like features. Some of them may have been difficult to put into words. Recognition of patterns and shapes takes place, after all, in the right side of our brains, on the opposite side of the centers that process verbal logic. "After considerable experience," Charles Darwin admitted in his treatise on barnacles, "when numerous varieties of a species have been carefully examined, the eye acquires a sort of instinctive knowledge, by which it can recognize the species, though the character cannot be defined by language." By 1924, Dart may have examined enough primate brain casts to acquire an element of that kind of "instinctive knowledge."

Then there's the matter of the foramen magnum—the opening through which the spinal cord enters the skull. In apes, that opening is situated toward the back of the cranium, reflecting the fact that the backbones of animals who walk on four limbs enter their skulls toward the rear. In humans, who walk upright, the backbone enters the skull from the bottom. Although the portion of the skull that had covered the Taung brain cast was almost entirely missing— destroyed, probably, in the blast that had detached the skull from the limestone that had contained it—one piece of it remained. A fragment of one of the four bones that are fused together to form the foramen magnum, the right exoccipital, was still attached to the bottom of the endocast like a scrap of postage stamp stuck to a parcel wrapping.

This enabled Dart to make a firm determination that the Taung being's foramen magnum was situated forward of the typical ape position. That was obviously a very important indicator that the species he'd identified had walked upright, but for Dart it only confirmed what his first sight of the brain cast had already told him. "I was convinced," he wrote, "from the earliest period of my investigations that these creatures had placed great reliance on their feet for walking and running and that, consequently, their hands must have been freed for other tasks. *This was implicit in the globular form of the skull which was obviously balanced on a more vertically placed type of backbone than that of a gorilla or chimpanzee*" (the italics are mine). We're thrown back, therefore, on the shape or "look" of the Taung endocast. That shape hadn't only convinced Dart that the Taung species was a human-like primate—it had also suggested to him that it had walked upright on two legs.

Without an understanding of Dart's exceptional exposure to comparative brain anatomy, one might well be skeptical—as his long-standing opponent Lord Solly Zuckerman was—of "the fossil ape-like skull which, presumably by divine guidance, Dart immediately recognized as the missing link." Dart's old chief, Grafton Elliot

Smith, had a better understanding of the level of experience and training Dart brought to bear on the Taung skull. "It was," Elliot Smith declared, "a happy circumstance that such a specimen fell into ... [Dart's]... hands, because he is one of, at the most, three or four men in the world who have the experience of investigating such material and appreciating its real meaning."

In the weeks to come, Dart would laboriously free the Taung primate's facial bones from their limestone casing. This would reveal it to be a child whose permanent teeth were just starting to erupt. It would also allow the human-like nature of the child to be established on criteria that were easier to demonstrate objectively: an upwardly rising forehead; a lack of ape-like brow ridges; a jaw that didn't protrude as far forward as those of apes; relatively broad cheek-teeth whose cusps were lower than those of apes; and canines and incisors that were smaller than those of apes. All this sounds settled and obvious today, after the discovery of numerous "post-cranial" or "below the skull" bones of "australopithecine" hominids has confirmed Dart's conclusions. From the perspective of 1925, however, Dart's announcement that the Taung primate had been a human-like, upright-walking being took extraordinary levels of both expertise and courage.

Anxious about the imminent arrival of the bridal party at their house, Dora Dart tried twice to call her husband away from the boxes of fossils that had been sent from Taung. Dart's mind was, however, too occupied with the implications of what he was seeing to hear her. He was trying to absorb the fact that he may just have identified "one of the most significant finds made in the history of anthropology."

> Darwin's largely discredited theory that man's early progenitors probably lived in Africa came back to me. Was I to be the instrument by which his "missing link" was to be found?

These pleasant daydreams were interrupted by the bridegroom himself tugging at my sleeve.

"My God, Ray," he said, striving to keep the nervous urgency out of his voice. "You've got to finish dressing immediately—or I'll have to find another best man."

* * *

The Taung skull reached Dart toward the end of November of 1924. In about forty days he'd freed the face from its limestone cover, had photographs taken, and published a description of the find in the February 7, 1925 issue of *Nature*. Dart was cautious. He didn't claim that the Taung child was a member of the human family, but placed it, instead, in a half-ape, half-human family reminiscent of Dubois's "Pithecanthropoidae," for which Dart used the name *"Homo-simiadae."*

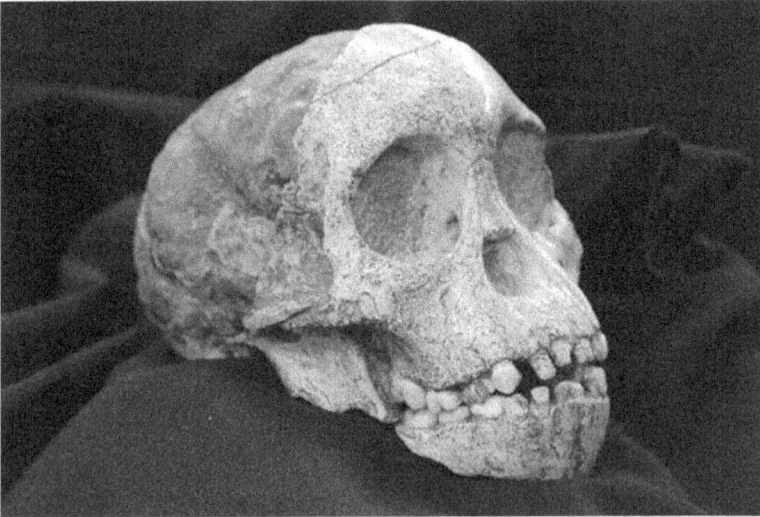

Reproduced with the kind permission of Bernhard Zipfel, Curator of Collections and the Evolutionary Studies Institute, University of the Witwatersrand, Johannesburg, and creator of this image of the Taung child.

Nature's editor elicited responses to Dart's report from four of Britain's leading physical anthropologists and published them in the next edition of the journal which appeared on February 14, 1925. Not one of those responses supported Dart's contention that the Taung infant was a human-like, upright-walking being. The consensus was, rather, that it belonged, as Arthur Keith put it, "in the same group or sub-family as the chimpanzee or gorilla." Even Elliot Smith—who would later be won over by his protégé's arguments—cautioned that its seemingly human-like features were "not unknown in the young of the giant anthropoids and even the adult gibbon." Arthur Smith Woodward—who had, in the previous year, become "Sir Arthur"—directed all the skepticism that he should have used on "Piltdown man" at Dart's find. He was unimpressed by the human-like narrowness of the Taung endocast: "...the Taungs [sic] skull lacks a brain-case, so the amount and direction of distortion of the specimen cannot be determined." "It is," he concluded, "premature to express any opinion as to whether the direct ancestors of man are to be sought in Asia or Africa. The new fossil from Africa certainly has little bearing on the question." Woodward expressed regret, too, that Dart had "...chosen so barbarous a (Latin-Greek) name for it as *Australopithecus*."

The full name that Dart had given to the Taung species was *Australopithecus africanus*. The "australo" part of the generic name was Latin for "southern" while "pithecus" was Greek for "ape." In our time, when a beetle can be called *Utu brutus*, a braconid wasp *Agra vation*, and a horsefly, *Chrysops balzaphire*, it's hard to understand how a scientific name could have been disapproved on the grounds that it combined Latin with Greek. Woodward was, however, not the only one to get huffy about it. "It is generally felt," an anonymous commentator in *Nature* wrote, "that the name *Australopithecus* is an unpleasing hybrid as well as etymologically incorrect." "If you want to join in a game," a British Museum scientist named F. A. Bather lectured Dart in *Nature*, "you must first learn the rules."

Undaunted by the less-than-enthusiastic British reaction to his find, Dart had plaster casts of the Taung skull made, painted to resemble the original. These were displayed in the South African pavilion at the British Empire Exhibition at Wembley in the summer of 1925. On a chart accompanying them, Dart suggested that his *Australopithecus* was the ancestor of Dubois's *Pithecanthropus*. The exhibit was set out under a banner that read AFRICA: THE CRADLE OF HUMANITY. That label has the easy familiarity of a cliché today, but it was a shocking one in 1925. It was a serious challenge, too, to the men who had built their careers and reputations on the proposition that the earliest-known ancestor of *Homo sapiens* had been found at Piltdown in England.

Keith—who'd also become "Sir Arthur" since the "discovery" of *Eoanthropus*—reacted to the Wembley exhibit by announcing that Dart's claims were "preposterous." Keith was the Pasha now—the Nabob of the anthropological community. At least within the British Empire, a word from him could make or break a young scientist's reputation. When the British Association for the Advancement of Science met at Leeds under his chairmanship in 1927, Keith pointedly omitted any reference to the Taung skull, emphasizing, instead, the importance of Piltdown man and Dubois's *Pithecanthropus*.

Keith was not so reluctant, however, to discuss Taung in print. Working from casts, he analyzed Dart's find at great length. In his 1931 *New Discoveries Relating to the Antiquity of Man*, he came to the unsurprising conclusion that it could not represent a human ancestor. A short while later, the Royal Society (of which Keith was now an influential member) refused to publish a 300-page monograph written by Dart on the Taung child—ironically because Keith had already dealt with the topic adequately in his *New Discoveries*. This was like refusing a defense attorney the right to speak at a trial on the grounds that the prosecutor had already explained the situation to the jury clearly enough. Dart's monograph remained unpublished and remains so to this day.

Although Dart was able to divert his attention back to his teaching duties and continue to help build up the medical school in Johannesburg, the disparagement of his scientific competence by the leaders of his profession put him under tremendous pressure. That pressure may have contributed to the breakup of his marriage in 1933. It may also have had something to do with the fact that he suffered two episodes that have been described as "breakdowns" in the long interval between the rejection of his view of the Taung skull in the mid-1920s and the final vindication of that view some twenty years later.

The first tangible element of that vindication came along in 1936 when Dart's indefatigable Scottish-South African champion Robert Broom discovered a second australopithecine—an adult specimen this time—in a karst cave just north of Johannesburg called Sterkfontein. American interest in the South African discoveries grew, and then, in the early forties, ripened into full agreement with Dart's views.

In 1947, Keith himself—now eighty-one years old—admitted on the pages of *Nature* that *Australopithecus* was on, or very close to, the line of human ancestry, writing that "I am now convinced on the evidence submitted by Dr. Robert Broome that Professor Dart was right and I was wrong." That admission might have seemed like a magnanimous one, but Keith was, in fact, backing away from Piltdown at the last possible moment. By 1949, Kenneth Oakley was to prove, by measuring the fluorine content of their organic specimens, that the Piltdown materials were much younger than they'd been assumed to have been. In 1953, Weiner, Oakley and Le Gros Clark were to show that the apparent association of those materials, and their presentation to the scientific community as a series of genuine finds, were products of the elaborate and carefully planned fraud discussed in the previous chapter.

How human was *Australopithecus africanus*? Would individuals belonging to that species have been able, for instance, to talk to one another? Dart had decided, as we've seen, that the region of *Australopithecus*'s cerebral cortex that was situated between, and adjacent to, the parallel and lunar sulci—the so-called parieto-temporal complex—was, compared to the same region in apes, significantly enlarged. Combining this insight with what he knew about the sensations, abilities and functions mediated by that region in *Homo sapiens*, Dart felt that he could hazard some guesses about the mental abilities of the human-like little creatures he'd discovered:

> They possessed to a degree unappreciated by living anthropoids the use of their hands and ears and the consequent faculty of associating with the colour, form and general appearance of objects, their weight, texture, resilience and flexibility, as well as the significance of sounds emitted by them. In other words, their eyes saw, their ears heard, and their hands handled objects with greater meaning and to fuller purpose than the corresponding organs in recent apes. They had laid down the foundations of that discriminate knowledge of the appearance, feeling and sound of things that was a necessary milestone in the acquisition of articulate speech.

Until recently, few anthropologists were receptive to the idea that the australopithecines could have "laid down the foundations" of linguistic communication. Many would have agreed, instead, with a frequently quoted 1982 statement by the archaeologist Randall White, that there was, before a 100,000 years ago, "...a total absence of anything that modern humans would recognize as language." As we'll see in the next few chapters, however, the idea that sophisticated human abilities (like the ability to invent new technologies and processes and the ability to hunt larger animals) appeared relatively recently and suddenly in a "great leap forward" is now being abandoned. In his *Language Instinct*, Stephen Pinker argues the thinkable in regard to our ability to communicate linguistically: that it may have appeared in the gradual way in which other human abilities evolved, and that its roots may well go down to our australopithecine forebears.

Dart was also convinced that *Australopithecus* could hunt animals larger than itself, a proposition that we'll discuss at length in the coming chapters. He concluded, in addition—on the basis of seemingly burned bones found at Makapansgat—that at least one of the australopithecine species could control fire. The last of these conclusions was clearly wrong—no corroborating evidence of australopithecine fire-use has ever come to light, and it was later proved that bones on which that conclusion rested had been blackened by oxides of manganese and iron rather than by fire.

* * *

I haven't tried to hide my admiration for the mixture of intuition, courage and hard intellect that Dart needed to establish and announce the fact that the Taung child represented an upright-walking being endowed with some of the physical and psychological attributes of our species. We must now look, however, at an area in which Dart's intellectual daring may have pushed him too far. That area concerns the tools or weapons that *Australopithecus* might (or might not) have used to kill and/or butcher other animals.

The sites where the South African australopithecines were found have, not surprisingly, an enormous quantity of non-hominid bones. Dart analyzed tens of thousands of those remains from Makapansgat and came to the conclusion that australopithecines had used them, in both modified and unmodified form, to kill and/ or cut up other animals. In bringing the bones to the cave, the australopithecines had, Dart believed, selected those that were usable as tools or weapons. This would explain, he argued, why gazelle horn-cores and the mandibles or lower jawbones of both antelopes and carnivores were so common in the Makapansgat assemblage. The former were used, as Dart saw it, as daggers, while the latter served—with their teeth in place—as cutters. It would explain, too, why antelope arm-bones or humeri were so common in the cave, and, indeed, why the distal or "lower" ends of those humeri were ten times as common as their proximal ends: because the distal ends were hard, Dart reasoned, they made better club-heads than the soft "cancellated" bone of proximal ends. The proximal ends had presumably been removed, therefore, and discarded outside the cave.

Moving beyond the speculative into the fanciful, Dart suggested that the absence of tails in Makapansgat cave "was probably due to their use as signals and whips in hunting outside the cavern." Dart, who had a predilection for what his successor Phillip Tobias called "gargantuan, sesquipedalian, Brobdignagian" words, named this presumed bone-, tooth-, and horn-using activity an "Osteodontokeratic Culture" (osteo = bone, donto = tooth, keratic = horn).

Dart didn't just see the australopithecines as hunters—he thought that they were, in addition, cannibals: "*Australopithecus* lived," he wrote, "a grim life." "He ruthlessly killed fellow australopithecines and fed upon them as he would upon any other beast, young or old." This conclusion was based on the fact there were fractures in some of the australopithecine skulls and jaws found in South Africa in the first half of the twentieth century. Dart was, as we'll see, simply too eager to see those fractures as evidence that the owners of the bones in question had been clubbed to death by their fellows. (The scene in *2001: A Space Odyssey* in which a human-like ape beats another to death with a long bone is pure Dart.)

Because of this supposed proclivity for cannibalism on the part of *Australopithecus*—and because of the fact that it was thought to have killed and eaten other animals—Dart regarded it, and its presumed descendant *Homo sapiens*, as uniquely *cruel* beings.

The loathsome cruelty of mankind to man is the inescapable product of his bloodlust; this differentiative human characteristic is explicable only in terms of man's carnivorous and cannibalistic origin. Charles Darwin had acknowledged that humans could be cruel, but he'd laid much of the blame for that on superstition. The fact that we no longer sacrifice humans to bloodthirsty gods, or subject them to trial by ordeal shows, Darwin thought, "what an infinite debt of gratitude we owe to the improvement of our reason, to science, and to our accumulated knowledge."

Perhaps because Dart had been exposed, as an army doctor, to some of the grim effects of the First World War, he was less optimistic about science and human reason. "Darwin could not guess," Dart wrote (commenting on the former's "debt of gratitude" phrase), "that within a century science would give birth to poison gases, wholesale human slaughter and atomic obliteration." The reason why Darwin didn't (in Dart's opinion) realize how deeply cruelty and aggression were rooted in human nature was that Darwin failed (as Dart saw it) to deduce "that man had inherited those qualities from his predacious ancestry."

As the Cold War climaxed in the Cuban missile crisis of 1961, the American playwright Robert Ardrey popularized Dart's ideas in a book titled *African Genesis*. Like *Anna Karenina* and *A Tale of Two Cities*, Ardrey's book starts with the kind of passage that sticks in people's memories:

Not in innocence, and not in Asia, was mankind born. The home of our fathers was that African highland reaching north from the Cape to the Lakes of the Nile. Here we came about—slowly, ever so slowly—on a sky-swept savanna glowing with menace.

The idea that humans were uniquely aggressive and cruel made sense to a reading public that was facing a real possibility of nuclear war. Ardrey told his readers that humans are descended from apes who were "armed killers." Their endless competition to develop superior weapons is, he concluded, "a genetic necessity." "Our first hominid ancestors," Colin Tudge wrote in his 1997 *Time before History,* "succeeded, probably, by being at times extremely unpleasant—to a degree which, among animals only modern chimpanzees achieve." This view of our species still persists.

Are we really to believe, though, that humans (and chimpanzees) are uniquely "unpleasant" because they commit "murders" and take part in "wars"? A great many other species, vertebrate and invertebrate, become involved in acts of interpersonal and intergroup aggression that result in violent death. Looking back on this Dart-Ardrey view of human nature from the year 2021, I have to confess that I can't see why any organism should be regarded as uniquely aggressive and cruel because it kills members of its own or other species.

In support of his idea that predation by hominids had led to the development of an extraordinary, aberrant level of aggression, Dart also pointed out that humans have "either decimated and eradicated the earth's animals, or led them as domesticated pets to his slaughterhouses." Even this seemingly telling point—that humans have exterminated other species on a large scale—does not, however, compel us to accept Dart's view of human nature. I'll argue presently that the damage humans are doing to the biosphere is the result of the unprecedented mental power of our species rather than a supposedly abnormal level of cruelty and aggression.

* * *

So much—for the time being anyway—about the psychological, ecological and ethical conclusions Dart drew from his osteodontokeratic or "ODK" theory. How has that theory itself fared in the half-century or so since Dart formulated it?

In 1965 C. K. "Bob" Brain, then a newly appointed paleontologist at the Transvaal Museum in South Africa, started analyzing the fossils in Swartkrans Cave. Increasing acquaintance with this and other bone assemblages brought Brain to the realization that the skeletal disproportion in Makapansgat that had been caused, as Dart saw it, by hominin selection was the result, instead, of the simple fact that some kinds of bone survive destructive processes better than other kinds. Some years previously Brain had already found, in a sample of modern goat bones discarded by Nama herders and then chewed on by their dogs, that eighty-seven distal humerus pieces had survived but not a single proximal end. The answer lay in the fact that the hard bone on the distal ends had survived the chewing, trampling and weathering they had been subjected to, while the fragile, spongy proximal ends were easily destroyed for the nutrients they contained. When Brain told Dart that the overall survival of the goat skeletal parts he'd collected in Namibia closely mirrored the survival proportions of the Makapansgat material, Dart was at first, Brain relates, "taken aback and perplexed."

> But after this initial dismay, he became increasingly enthusiastic, saying "this is wonderful—now at last we are getting closer to the truth." Rather than condemning this young upstart who was upsetting his cherished concept, Dart nominated me for an award. I realized then that he was much more interested in the subject of his investigations than in his position relative to them and that he was one of those rare individuals with true generosity of spirit, particularly in this rather emotional field of paleoanthropology.

In his 1981 *Hunters or the Hunted?—An Introduction to African Cave Taphonomy*, Brain demonstrated convincingly that most of the bones at Makapansgat and several other South African caves of the same kind, in which early hominid remains had been found, were accumulated by non-hominid agencies such as water, gravity, and non-human predators, and that those bones had, in many

cases, been modified by non-hominin carnivores and by porcupines. The fractures of australopithecine crania which Dart had
regarded as evidence of blows delivered by antelope leg-bone clubs
were, Brain showed, more likely to have been caused by the bites of
non-hominin predators such as leopards. This critique led to a total
rejection by the paleoanthropological community of Dart's notion
that the australopithecines were hunters and tool-users.

More recent discoveries have, however, reinforced the fact that
Dart's intuition about australopithecine tool-use was pointing
him in the right direction. In 2010, two bones dated to 3.4 million
years ago—one from the right rib of a buffalo-sized ungulate, the
other from the leg of a smaller one—with cut-marks probably made
with stone tools, were found in Ethiopia's Afar Region. The understandable skepticism that greeted this find gave way five years
later, when the first direct evidence of deliberately manufactured
stone tools showed up in 3.3-million-years-ago deposits at a site
named Lomekwi 3 on the west side of Lake Turkana in Northern
Kenya. Astonishingly, these indications of stone-tool use *predate*
the appearance of *A. africanus* and reach back into the time of *A.
afarensis*—Lucy's species.

Dart's ODK theory has itself turned out to contain a grain of truth:
by the 1980s Bob Brain had identified wear-marks on the ends
of about sixty bone specimens from Swartkrans's three older
Members (consisting mainly of the long bones and the horn-cores
of antelopes, but including, too, the left lower jawbone of a three-
toed horse) that demonstrate clearly that those bones had been
used as digging implements.

Brain showed a selection of these implements to Dart, who was by
this time close to ninety and almost blind. "Brain," the latter replied,
after feeling their shape with his fingers, "I told you a long time ago
that *Australopithecus* made bone tools, but you didn't believe me.
Now, what do you suppose these were used for?"

Brain replied that they were almost certainly used for digging in the ground.

"That," Dart replied, "is the most unromantic suggestion that I have heard in my life." Despite the fact that he was well-disposed toward Brain's work, the old man clearly hadn't let go of the melodramatic aspects of his ODK theory. Picking up one of the sharpest bone tools on the table, he pressed the end into Brain's ribs. "I could run you through with this," he said.

Drama and hype had long been elements of Dart's style, and nobody was going to change that. An earlier suggestion, made privately to him by his successor in the chair of anatomy at the University of the Witwatersrand, Phillip Tobias, that his tendency to overstate his arguments was putting people off, was dismissed in good humor:

> He looked at me, not unkindly, but bridling a bit, and said, "Phillip, I have to do it this way, with such a new and revolutionary concept." He warmed a little: "If you don't give the [expletive] 200 per cent, they [expletive]-well won't believe the half of it." The frown gave way to a sudden warm smile and chuckle.

* * *

On a cold July day in the Southern Hemisphere winter of 1932, my maternal grandfather, Thomas Frederik Dreyer, the Professor of Zoology and Geology at Grey University College in Bloemfontein, South Africa, was excavating fossilized material from one of the springs at Florisbad, near Bloemfontein in South Africa's central interior. An assistant by the name of Wildeboer Venter, working in the "eye" of this spring, was passing fragments consisting mainly of pieces of black, gummy wood and mammalian remains up to Dreyer. Between inspecting these items, Dreyer was helping the laborers he'd hired to bail water out of the eye.

The men had been working for hours, and everyone was wet and cold. Dreyer forgot his discomfort abruptly, however, when he recognized one of the items that Venter had passed up to him as a last upper human molar. The next item to emerge from the water-logged sand of the "eye" was the roof of a human skull. A few days later, when the material taken from the eye was more carefully sorted, Dreyer found pieces of the face. The walls of that skull were thicker than those of modern skulls and the brow-ridges, somewhat heavier. Those attributes spoke of a somewhat archaic being, but the skull's general appearance was much closer to that of a modern human than it was to either Kabwe (*Homo rhodesiensis* aka *heidelbergensis*) or to the Neanderthals.

The fact that the partial skull that emerged from Florisbad looked "modern" did not mean, however, that it was recent. The spring complex in which it had been found had also yielded the remains of

Reproduced with the kind permission of Lloyd Rossouw, Florisbad Quaternary Research, National Museum, South Africa.

extinct equine and buffalo species. That, combined with the skull's archaic features, suggested to Dreyer that the human it represented must be an ancient one—a contemporary, as he saw it, of the Neanderthals. At that time many people assumed that Neanderthals were the ancestors of modern humans and that the Florisbad human must, therefore, have (a) emerged *after* the Neanderthals, or (b) actually have *been* a Neanderthal.

"A prominent South African anthropologist when he saw the skull strongly advised me to describe it as Neanderthal—if I would have it receive due attention..." Dreyer wrote in a memorandum on file at the National Museum. "This," the note continues, "I could not see my way clear to do." One look at the Florisbad cranium is enough to establish that it resembles the skulls of present-day humans far more closely than those of Neanderthals.

Because it didn't fit into the then-current scheme of human evolution, the Florisbad find did not receive the "due attention" that Dreyer's colleague had spoken about. "[I]ts very peculiar features," Dreyer wrote in 1947, "have not gained for it the measure of attention that it certainly deserves." As Dreyer saw it, the problem lay in "the inability of the adherents of the Neanderthal theory [i.e., the Neanderthals are our ancestors theory] to envisage a primitive human form with a *H. sapiens* form of face."

* * *

By the early 1950s Raymond Dart, based in Johannesburg, some 250 miles north of the Bloemfontein/Florisbad area, had, as we've just seen, long since been recognized for his 1924 identification of the Taung child as a hominin. That vindication stood in sharp contrast to the continuing lack of enthusiasm and comprehension with which the scientific community viewed Dreyer's find. The fact is that, until recently, nobody possessed enough information to understand how the Florisbad skull fit into the puzzle of human evolution. Dreyer himself understood only that the "primitive human form with a H. sapiens [sic.] form of face" he'd discovered must be a significant piece in that puzzle.

I was very close to my grandfather as a child, so I find it hard to write, now, that it is no credit to Dreyer that he was one of the majority of scientists who initially rejected Dart's claim that *Australopithecus* was a member of the human family. Although he and Dart had not, probably for that reason, enjoyed a warm relationship, Dart persuaded his colleagues at the Witwatersrand University to confer an honorary doctorate on Dreyer in 1954, in recognition of the latter's archaeological and paleontological achievements—not least of which being the finding and evaluation of the Florisbad skull. Any negative feelings that may have remained melted away with Dreyer's trip to Johannesburg for the conferral of that award.

Some months later, Dart and a number of other Witwatersrand anatomists and anthropologists traveled from Johannesburg to Bloemfontein to visit with Dreyer. Phillip Tobias, later to be appointed as Dart's successor, remembered the warmth of that meeting. He also remembered feeling nervous about the fact that Dreyer was holding the precious Florisbad cranium in an unsteady grasp while addressing the group. Dreyer had gotten out of a sickbed for that visit and was to die later that year.

Dreyer had named his find *Homo helmei*, for Captain C. Egerton Helme, the son of a Lancashire industrialist who'd emigrated to South Africa for health reasons after active service in the First World War. Helme had provided funding for my grandfather's Florisbad excavations and assisted him personally with some of them. Until recently, the *helmei* name was seldom used; the Florisbad human was seen, instead, as a member of *Homo sapiens*. To distinguish it from "archaic" *Homo sapiens*—that is, Kabwe and its contemporaries—it was classified as an "early modern" member of our species. By the 1990s a consensus was developing, however, that the *Homo sapiens* taxon had become an overstuffed cupboard. Neanderthals—formerly *Homo sapiens neanderthalensis*—were taken out of it and returned to their own specific category, *Homo neanderthalensis*, which had been assigned to them by William King in 1864. Kabwe's original *Homo rhodesiensis* name was also restored, and people started using the *helmei* name Dreyer had given to his Florisbad find.

During Dreyer's lifetime, it was thought that the Florisbad skull could be something like 40,000 years old (probably because the radiocarbon testing that had been done on it could measure that far, but no further, into the past). Estimates of its age gradually crept upward during the second half of the twentieth century. Sixty thousand years was mentioned, and then 100,000 and more. In 1991 the late James Brink, then-Director of the Florisbad Quaternary Institute, and curator of the skull, approached Rainer Grün of the Quaternary Dating Research Centre of the Australian National University in Canberra to date the Florisbad site and the fossils found there. Dates ranging from 100,000 years to 400,000 years BP were obtained by a mix of optically stimulated luminescence (OSL) tests of quartz crystals from the site, and electron spin resonance (ESR) tests of the tooth enamel of fauna found there. In 1994 Brink and Grün decided to remove two fragments of enamel from the single human tooth found with the Florisbad skull and do direct ESR tests on each of them. Those tests revealed that the human whose remains Dreyer had unearthed had lived some 259,000 years BP. Now, finally, the full significance of Dreyer's find began to emerge. If Dart's Taung child had demonstrated the unexpected and even shocking fact that the human family had evolved on the "primitive" African continent, then the Florisbad human was the bearer of an equally startling message: millions of years after the origin of the hominin family, its last-surviving and most remarkable member, *Homo sapiens*, also had its origin in Africa.

This realization has since been confirmed by the discovery, elsewhere on that continent, of several other hominin specimens closely resembling Dreyer's *Homo helmei*, that span the relevant time frame.

Chapter 10

Traveling Through Time in a Cluster of Species

All the way from its origins around 7 million years ago up until just over 2 million years ago, the hominin family seems to have consisted *exclusively* of small species. Like baboons, early hominins were "sexually dimorphous" creatures. The males weighed around ninety pounds and grew about five feet tall, while the females weighed only about fifty pounds and stood four feet high or less. We would have been struck not only by the diminutive stature of those "australopithecine" hominins but also by the unfamiliar proportions of their bodies. Whereas our chests taper down into relatively narrow waists, the thickish midsections of the australopithecines tapered upward into a relatively narrow chest and shoulders. The powerful, sinewy arms that hung from those narrow shoulders would have seemed too long for their owner's short thighs.

* * *

In the classic view of human evolution, hominins split off from their ape-like ancestors in the process of leaving the forests in which they had evolved and starting to live on the savanna. This view became unfashionable after the tree-climbing adaptations of the australo-pithecines we've just been talking about became apparent. More evidence was aligned against the "we moved on to the savanna" hypothesis when fragments of forest plants like lianas were found in the same deposits as *Australopithecus* fossils, along with the remains of forest animals like colobus monkeys, the woodland baboon *Theropithecus brumpti*, tragelephine or "bushbuck-type" antelopes and reduncinae or "waterbuck-type" antelopes.

For a while it looked like the savanna hypothesis would have to be scrapped completely, but australopithecine remains are not *exclusively* associated with those of "moist-woodland" mammals. They're found, too, in association with those of early impalas, that lived in relatively dry bush country, the savanna baboon *Theropithecus dartii* and the "alcephaline" grassland-dwellers, wildebeests and hartebeests. It's beginning to look, therefore, as if the australopithecines might have lived in a mosaic of forest and grassland.

Back in the middle Miocene, around 15 million years ago, famous hominin sites like the middle section of the Awash River in Ethiopia, Olduvai Gorge in Tanzania and the Sterkfontein Valley in South Africa were probably covered, like most parts of Africa were at this time, by dense forest. Beginning about 10 million years ago, a cooling and drying process started opening up the forest cover in those and other parts of Africa. That "opening-up process" didn't immediately create grasslands as dry or as extensive as those presently found in Africa. It seems, instead, to have created regions of *relatively* diminished rainfall in which forest was transformed into the kind of heavily wooded savannas that are found today on the margins of the Congo's rainforests.

The australopithecines were probably dependent on both the grassland and the forest components of those Congo-like savannas. The abundance of trees on the Miocene and early Pliocene savannas may have provided them with an essential refuge from dangerous animals. There were, as we'll see in the next chapter, no less than six kinds of big cat in late-Pliocene Africa, as well as four or five kinds of hyena, ranging from the lion-sized *Pachycrocuta* to the cheetah-like *Chasmaporthetes*. Some of those carnivores probably killed hominids for food on a fairly regular basis. Modern-day leopards are, Jean Dorst and Pierre Dandelot's *Field Guide* tells us, "...particularly partial to monkeys and baboons." During the late Pliocene and early Pleistocene, that predilection for primates extended to australopithecines: predation by large cats such as leopards and sabertooths is thought to be an important factor contributing to the accumulation of hominin remains as fossils at Swartkrans.

The increases in both size and technological proficiency that humans have undergone in the last 2 million years have made them somewhat less vulnerable, but leopards still kill humans for food occasionally in both Africa and Asia. Walking through Berg-en-Dal Rest Camp in the southwestern end of the Kruger Park, I noticed a memorial stone with fresh flowers on it which read:

> In memory of our beloved son and brother Charles Aldridge Swart 18/8/73—21/8/98, student ranger at Berg-en-Dal who was killed on duty by a leopard...

I learned subsequently that the animal had attacked Swart in a completely unprovoked way and that it had eaten part of his body after killing him. Since Swart's death, at least three other people—the ten-year-old son of a park employee, a fifty-eight-year-old woman who worked at Skukuza, and, in June of 2019, a thirty-month-old toddler—have been killed by leopards in the Kruger Park.

According to official records, the man-eating leopard of Rudraprayag in Uttarakhand, India, killed and ate 125 people, but Jim Corbett, who shot the animal in 1926, thought that the number of deaths was probably much higher due to unreported kills and deaths due to injuries sustained in attacks.

Corbett, who became famous for hunting down man-eating tigers and leopards in India in the early twentieth century, defended India's big cats by arguing that man-eating is an "unnatural" activity for them, indulged in mainly by injured or old animals who can no longer hunt "natural" prey. I suspect, however, that man-eating only seems abnormal in our time because we've dramatically lessened the numbers of big cats and because, under normal circumstances, we protect ourselves so effectively against the remaining ones. Cats are intelligent animals who are well aware of the power that humans usually have at their disposal. Lions have never attacked the tourists who participate in the Kruger Park's organized "bush walks," even though they've been encountering those participants, together with their armed escorts, on a regular basis for decades. Lions regularly encounter *unarmed* people walking on foot in Southern Africa—I have had several such meetings myself—and do not, in the vast majority of cases, kill such people either.

In the early 1990s, however, when refugees from war and famine in Mozambique were crossing into South Africa through Kruger's wilderness, lions quickly realized that those particular humans weren't part of the defense and retaliatory network that protect the other humans in the area. Those unarmed refugees were, for one thing, spending the night in the veld. As a result, the cats killed and ate what was, from our species' point of view, a horrifyingly large number of them. Kruger lions have also been reported, less horrifyingly perhaps, to have killed and eaten a rhinoceros poacher.

Between 1932 and 1947, lions killed 1,500 people in the Njombe District in southern Tanzania. A more recent report tells us that, since 1990, over 600 people had been killed by lions in that country. Reports of relatively large numbers of humans being killed by lions have also come out of Ethiopia recently and it's highly likely that similar things are happening in other parts of Africa outside the view of the media. It's reported, too, that the lions who are doing this killing aren't predominantly old or injured animals. For well over 99 percent of their evolutionary history, lions and leopards preyed on hominins in an entirely "natural" way. Our instinctual fears still reflect that state of affairs—our children still get genetically encoded messages that scary monsters can materialize out of the darkness.

Predators like big cats and giant hyenas weren't the only mammals that posed a threat to the afarensis australopithecines. The pig and baboon species that competed with early hominins for savanna foods would almost certainly have come into conflict with them from time to time. Initially, at any rate, those conflicts would have been one-sided affairs. The overwhelming size advantage that the biggest kinds of pigs and baboons had in the Pliocene, together with the formidable tusks of the former and fangs of the latter, would have given them the ability to shred an australopithecine in short order.

* * *

Perhaps because the ability to climb into a tree in a quick and agile way was an important defense against these formidable predators and competitors, the australopithecines were much better adapted to tree-climbing than modern humans are. The long, curved finger- and toe-bones of Lucy's species, *Australopithecus afarensis,* as well as its long, ape-like forearms and short legs, speak of a need to climb trees that persisted long after hominins started walking upright some 6 or 7 million years ago. That conclusion is bolstered by evidence suggesting that the still-unnamed 4.17-million-year-old "little foot" australopithecine from South Africa had opposable big toes, resembling the opposable thumbs on primate hands. That would have enabled its feet to *grasp* branches in the way apes' feet can, rather than just standing on them like our feet do.

There are claims for a slightly divergent toe and less well-developed arches in the Laetoli 3.6-million-year-old *Australopithecus afarensis* footprints and in the *A. africanus* bones of Stw 542 from Sterkfontein. However, closer analysis of both of these specimens indicates a fundamentally human pattern. If tree-climbing adaptations persisted, they were minor and do not undermine the fundamental fact that the feet of all hominins were dramatically reshaped by selection for bipedal performance at the expense of arboreal agility, showing the relative importance of these locomotor modes for the animals' survival and reproductive success.

The *smallness* of the australopithecines was among the most important of their tree-climbing adaptations. Gorillas aren't good tree climbers, despite the fact that their grasping feet and hands are well-suited to that task. That's so because they weigh, on average, almost two-and-a-half times as much as humans do. Gorillas can climb trees, but they do so, as Jean Dorst and Pierre Dandalot's *Field Guide to the Larger Mammals of Africa* tells us, "with great caution." Young gorillas climb trees far more easily than adults, while chimpanzees, who weigh, on average, about 70 percent as much as humans do, "climb trees with consummate agility and may occasionally jump to a nearby or lower tree..." Size is, therefore, of critical importance in the business of getting into trees and moving around in them efficiently.

Leopards, which weigh between 110 and 180 pounds, commonly protect themselves and their kills by climbing into trees that aren't accessible to 360- to 450-pound lions. Lions can and do make their way into relatively easy-to-climb trees, but the simple fact that they're so much bigger than leopards means that there's a wide range of trees that are not accessible to them, but that *are* accessible to leopards—even leopards hauling 70 pound gazelle carcasses up with them.

This "selective accessibility" of trees is a result of what we could call the "length-to-volume ratio." A four-foot-high cube is two-thirds as high as a six-foot-high cube, but its volume is only 29.93 percent of its six-foot counterpart. The mathematics of cubic volume works the same way for humans: a four-foot-tall human boy is two-thirds as tall as a six-foot man, but the boy's volume is just under a third of that of the six-footer. Growth tables confirm that four-foot-tall American boys (who reach that height at just over seven years of age) have a median weight of around fifty-two pounds—just less than a third of the weight of an average six-foot man. It is not, therefore, only their liking for play and adventure that make children better tree climbers than adults—a child who's fully two-thirds as tall as a given adult only has to haul about one-third of that adult's weight up into a tree. An australopithecine who was 80 percent as tall as an average modern human would only have had to haul about half of such a human's weight up a tree.

If the ability to climb trees in a quick and agile way was, in short, important to the survival of the australopithecines, it makes sense that natural selection would have kept their size within the chimpanzee range.

It's easy to imagine a hominin climbing a tree to escape from a large, angry pig, but it seems absurd to think of a member of the human family doing so to seek refuge from a baboon. If the hominin in question only weighed around seventy pounds, however, and the baboon was as big as a present-day gorilla (as at least one baboon species was in the early Pleistocene), then the logic of the length-to-volume ratio tells us that tree-climbing could have been an effective and sometimes, indeed, *easy* avenue of escape for the former. Seventy-pound hominins could, in fact, have escaped from 140-pound leopards on occasion by climbing quickly enough into the right kind of tree.

The proximity of trees would, therefore, have provided our Pliocene ancestors with what was often a dependable refuge from predators and non-hominin competitors. We can imagine australopithecines keeping a cautious eye out for climbable trees, therefore, as they walked across those heavily wooded Pliocene savannas. Personal experience informs me that hominins still do this when they're walking unarmed in bushveld inhabited by megaherbivores and big cats.

Walking across the veld during the day would have posed a moderate level of risk to the australopithecines, but sleeping on the ground at night—the prime hunting time of many of the big predators—would have been suicidal. (Raymond Dart named one of their variants *Australopithecus prometheus*, but there is no evidence of any kind that the australopithecines had the power to keep or make the fire that would keep the nocturnal predators at bay.) Seeking refuge in the rocky hills or "koppies" that dot the plains of eastern and southern Africa would have been a better option than sleeping on the open veld, but baboons also like to sleep on koppies. It seems possible, therefore, that some australopithecines might have constructed the kind of "nests" or "platforms" that allow modern-day chimpanzees to sleep in trees at night.

Australopithecines would also have climbed trees to get at fruit like the figs, sour plums, *Strychnos* "oranges" and marula "apricots" that are still abundant on the savanna at certain times of the year. But, even though the heavily forested Pliocene savannas must have produced a bigger variety of fruits than their more arid, present-day counterparts, fruit does not seem to have been the main source of food for the australopithecines. They seem to have specialized, instead, in finding and eating relatively hard, dry, fibrous grassland foods like seeds, nuts, roots, corms, tubers and bulbs. This is, at any rate, the conclusion that most people draw from the fact that the molar teeth of the australopithecines were broader than those of apes and covered, in addition, with a thicker layer of enamel. Apes, who live on fruits and other soft forest foods like shoots, vines and leaves, don't require heavy-duty molars of that kind.

Why would hominins have specialized in hard "grassland foods" rather than fruits? They were, I suspect, *obliged* to do so because the fruit niche was already occupied by their closest relatives, chimpanzees-to-be and other apes. Apes are regarded as somewhat pathetic creatures nowadays—animals that urgently require protection against our species. In the earlier stages of hominin development, however, they would probably have been formidable competitors in the business of living on forest food like fruits, shoots and leaves. Developing a way of eating and living that lay, for the most part, *outside* that niche, probably created our identity as hominins and allowed us to maintain it in the face of our anthropoid relatives.

As the writing of this book is being completed in 2021, the 6- to 7-million-year-old *Sahelanthropus tchadensis* appears to be the oldest hominin species yet discovered. One of several indications that *Sahelanthropus* is, in fact, probably a hominin, is the fact that its molar teeth are slightly broader than those of contemporary chimpanzees and covered, in addition, with slightly thicker enamel. By 4 million years ago, the molars of *Australopithecus anamensis*—a possible ancestor of *A. afarensis*—are already *considerably* broader than those of apes and covered with much thicker enamel. It seems likely, therefore, that hard, dry and fibrous foods—savanna foods—may have been part of the hominin diet from the start.

* * *

We'll probably never know how many species of australopithecine there were because primate populations may, as Alan Walker pointed out, *look* very different from each other as living beings, but have a very similar bone structure. Depending on who's doing the counting, there are, for instance, sixteen to eighteen species of long-tailed or "guenon" monkeys in Africa. If you look up "guenon monkeys" in Google Images, you'll see dramatic differences in the coloring and hair growth that make it easy to tell most of those species apart. As Walker pointed out, however, many of them would probably be indistinguishable on the basis of skeletal remains.

Even in the cases where there *were* skeletal differences between the various kinds of australopithecines, recognizing them in the fossilized fragments at our disposal can be a very subjective matter. Raymond Dart thought, for instance, that the australopith-

ecines found at South Africa's Makapansgat site, which he named
A. prometheus, were distinct enough from the *A. africanus* found
at Taung and Sterkfontein to be placed in a different species. The
scientific world hasn't recognized the former species, but Phillip
Tobias's analysis of the relevant specimens has brought him to the
conclusion that Dart's distinction is a valid one. In Tobias's view, the
Makapansgat australopithecines show affinities with *A. afarensis,*
whose remains have hitherto only been found near the Horn of
Africa, thousands of miles to the north.

Australopithecus afarensis was discovered in 1971 when a team
led by Don Johansen and Maurice Taieb found a large number
of hominin bones in the Hadar region of Ethiopia, including the
partial skeleton of the three-foot-tall individual that has become
known as "Lucy." As far as we can presently tell, *afarensis* existed
from 3.8 to just under 3 million years ago. (It probably coexisted,
therefore, at least for a short while, with *africanus,* which seems to
have existed from about 2.8 to 2.3 million years ago.)

In 1995 a team led by Michel Brunet named a new species called
Australopithecus bahrelghazali on the basis of a 3.3- to 3-million-
year-old jaw fragment that they'd found in Chad. Then, in 2001,
Meave Leakey described a 3.5-million-year-old hominin that she
and her colleagues regarded as being distinct from *afarensis.* They
assigned their find (which they'd excavated near the western shore
of Lake Turkana) to a genus of its own, calling it *Kenyanthropus
platyops* (Kenya human with a flat face).

In 1999 Leakey and her coworkers had described an older-than-*afarensis* species on the basis of fragmented remains found at both
Allia Bay on the eastern shore of Lake Turkana and at Kanapoi
on the lake's western side. Named *Australopithecus anamensis,* it
dates from between 4.2 to 3.9 million years ago. A partial tibia that
includes part of the knee-joint (and is very similar to tibias assigned
to *afarensis*) indicates clearly that *anamensis* walked upright. Other
features of its anatomy show, however, that it is a more ape-like

species than *afarensis*. Its teeth are arranged in the U-shaped arch typical of apes, for instance, rather than in the open parabola that characterizes later hominins. Its molars and premolars are, however, broad and thickly enameled enough to provide adequate confirmation of their owners' hominin status. *Anamensis* might (as Leakey herself has speculated) have been ancestral to *afarensis*.

Also in 1999, the uniquely complete skeleton of a still-unnamed contemporary of *anamensis* was pieced together in South Africa. The piecing-together process began when the British-South African paleontologist Ron Clark found hominin foot bones in a museum storage box labeled "cercopithecids." Following his suggestion to examine the wall of the Sterkfontein grotto at the place from which the stored remains had originally been dynamited, Clark's associates Steven Motsumi and Nkume Molefe managed to identify the location of the rest of the skeleton. This hominin, nicknamed "little foot," has been determined by the cosmogenic burial dating technique to be 3.67 million years old.

In 1994, Tim White and his colleagues discovered 4.4-million-year-old fragments of several apparent hominins in Ethiopia's Middle Awash district. On the basis of those fragments, they described a species called *Ardipithecus ramidus*. Searching at a nearby location in 1997, Yohannes Haile-Selassie, a member of Tim White's team, found the first in a series of remains of a still-older species that was named *Ardipithecus kedabba*. If *Ardipithecus* is in fact a hominin, the discovery of *kedabba* pushed the history of our family over the 5.3-million-year-ago boundary between the Pliocene and the Miocene.

Ardipithecus kedabba seems to have been a very ape-like species. The breadth of its molars, and the thickness of their enamel coating, are reported to be intermediate between the living apes and *Australopithecus*. A single toe-bone, a few hundred thousand years younger than the other remains ascribed to this species, resembles that of a bipedal hominin rather than that of an ape. (The toe-bones of clearly bipedal hominins like *africanus* and *afarensis* are much thicker and more robust than those of apes.) Was this very ape-like but apparently bipedal creature the last common ancestor ("LCA") of all the hominids? Its discoverers obviously thought it might be: "Ardi" means "ground" or "floor" in the local Afar language; "ramid" means root, while "kedabba" means "basal ancestor."

In 2000, however, Brian Pickford and Brigitte Senut announced the discovery in Western Kenya of a 6-million-year-old hominin they named *Orrorin tugenensis*. This name also implies a claim to the LCA position—"orrorin" means "original man" in the Tugen language spoken in the area where the finds were made. It may also have been intended to evoke the French word *aurore*—"dawn." Senut and Pickford claim that there's clear evidence that *Orrorin* was bipedal and that the enamel on its molars is thicker than that of *Ardipithecus*.

The history of the earliest australopithecines was made even more complicated and fascinating in June of 2002 with the announcement that a 6- to 7-million-year-old primate with hominin characteristics had been discovered more than a thousand miles west of the Rift Valley in west-central Chad by a team led by Michel Brunet (the discoverer of *Australopithecus bahrelghazali* that we talked about a few paragraphs ago). The discovery of this enormously ancient species—the *Sahelanthropus tchadensis* we talked about in connection with the diet and tooth structure of the earliest hominins—has cast doubt on the idea that hominin evolution was, in Yves Coppens's words, an "east side story"—that is, one whose action was restricted to the eastern side of Africa.

Sahelanthropus's foramen magnum is reported to be somewhat shifted to the front, but there are no other indications at the time of this writing that it was bipedal. If we haven't actually reached the time when hominins were quadrupedal with the discovery of Sahelanthropus, we could be close to it. Are we certain that there was a time when hominins were quadrupedal? The answer is "almost."

It's just possible that the chimp-human branch of primates had become bipedal before humans-to-be separated from chimps-to-be. If we assume this, however, we must also assume that chimps later re-evolved their quadrupedal way of walking. It's a more economical and generally far more likely assumption that hominins first split off from the chimpanzee branch and then became bipedal.

*　*　*

The word "australopithecine" doesn't denote a formal biological category like a genus or a sub-family. As I use it in this book, it's an informal designation, rather, for hominins whose small size, short legs and long, ape-like forearms made them more or less similar to Dart's *Australopithecus africanus*. Up to about 1.8 million years ago, when larger, differently proportioned hominins start to appear on the scene, all members of the human family could, according to this definition, be regarded as "australopithecines."

As the fossil evidence at our disposal already suggests, it's likely that the australopithecines branched out, relatively soon after they split off from the chimpanzee group, into several different species. Like many other mammalian groups, the australopithecines seem to have traveled through time in a cluster of species. Hartebeests, the grass-eating antelopes whose remains are sometimes found in association with those of hominins, exist in that kind of cluster. At its center are the "true" hartebeests: Coke's hartebeest, Lichtenstein's hartebeest, Swayne's hartebeest, Jackson's hartebeest, the tora hartebeest and the red hartebeest. On its outer margins, are "near-hartebeests": Hunter's antelope, the topi-tsessebe group and the blesboks.

The "true hartebeests" are very similar to each other. Except for Lichtenstein's hartebeest which lives in forested grassland, they are all open-savanna animals. One of the reasons hartebeests seem to have radiated into different species is the simple fact that they are spread right across Africa. The fact that australopithecines were also widely distributed through Africa and that the wooded savannas they occupied have been separated and fragmented repeatedly by wet-dry glacial cycles makes it understandable that they, too, would have split into a number of separate species or variants.

After about 3 million years ago—around at the time, therefore, that *Australopithecus afarensis* was fading from the scene and *Australopithecus africanus* was making its first appearance—the australopithecine cluster diverged into "gracile" and "robust" groups.

The "robusts" evolved massively built skulls, designed to anchor big muscles that powered outsized jaws and molar teeth. Those skulls were widened by enormous, protruding cheekbones and topped (in the males) with "sagittal crests"—ridges that run across the top of the skull in the same position that a rooster wears its comb. The possessors of this heavy-duty chewing apparatus do not, however, seem to have had larger or more robust bodies than those of the graciles. Some anthropologists place the robusts in the genus *Australopithecus*, while others consider them to be different enough to merit their own genus, which they would call *Paranthropus*.

The graciles had what we would regard as normal-sized cheekbones. Sagittal crests were either reduced or absent among them. Their skulls were thinner and their jaws, smaller. Still bigger than those of later humans, the cheek teeth of the graciles were also smaller than those of the robusts. The powerful chewing-apparatus of the robusts suggest that they were eating a coarser, more fibrous diet than that of the graciles. The graciles themselves ate what modern humans would consider to be a forbiddingly high-fiber diet, but it's thought that they had started processing their food outside their mouths to a greater degree than the robusts, cutting, scraping and/ or mashing it, perhaps, with crude tools before eating it. We know that Dart was overeager to identify horns and bits of broken bone as australopithecine tools. Is it possible, however, that *Australo-*

pithecus africanus and/or one or more of its contemporaries did in fact use tools of one kind or another to obtain and/or process their food? And could they, more specifically, have fashioned stone tools for those purposes? For years there was a gap between the latest australopithecines to be found, (*A. africanus*, which made its last appearance around 2.3 million years ago) and the earliest stone tools (found in a 1.8-million-year-old layer at Olduvai in northern Tanzania).

Stone tools of a relatively simple kind (cutting-flakes chipped off apple-sized "core stones") had been turning up in that 1.8-million-year-old Olduvai layer for decades. There was no clue, at first, as to the kind (or kinds) of hominin that had produced them.

Then, in 1960, Jonathan Leakey discovered fragments of a hominin skull in Olduvai's 1.8-million-year-old, tool-bearing layer. This skull, cataloged as OH 7, was found, after reconstruction, to have a capacity of 650 cc (vs. the approximately 500 cc typical of *Australopithecus africanus*). The walls of OH 7's skull were also somewhat thinner than those of *africanus* and its teeth, marginally smaller. A robust australopithecine named *Zinjanthropus* (now regarded as *Paranthropus*) *boisei* had been discovered by Mary Leakey in a contemporaneous or near-contemporaneous layer at Olduvai a year before. Louis Leakey assumed, however, that the species represented by OH 7, rather than *Zinjanthropus*, was the maker of the tools that had been found elsewhere in the 1.8-million-year-old layer. Because the tools in that layer were, in 1960, the oldest ones yet discovered, it was further assumed that OH 7's species had been the *first* one to make stone tools.

On the strength of the latter assumption, Raymond Dart suggested to Louis Leakey that OH 7 be called *Homo habilis*. "The specific name," Leakey, Tobias and Napier explained in their announcement of the new species, "is taken from the Latin meaning 'able, handy, mentally skillful, vigorous.'" Placing OH 7 in the genus *Homo* was the daring kind of move that appealed to iconoclasts like Dart and Leakey. In doing so they were, in effect, giving the middle finger to an anthropological world that had for decades ignored or disparaged their ideas and discoveries.

Understandably, this provoked what Richard Leakey described, in his *Origins of Humankind*, as "a storm of objections." Many scientists thought that OH 7's species should have been called *Australopithecus habilis*. A subsequent find that included a substantial amount of "post-cranial" material, the so-called "dik-dik hominin" (OH 62) unearthed at Olduvai in 1986, confirmed that *habilis* was, in fact, typically australopithecine as regards both size and body proportions.

Leakey and Dart seem to have reasoned this way in placing OH 7's species in the genus *Homo*: even if the changes in morphology or body-shape displayed by OH 7 were relatively small ones, the momentous new behavior that this being seemed to have acquired— making stone tools—was significant enough to warrant putting it in a different genus and, moreover, into the human genus. Just as Eugène Dubois probably wouldn't have called his 1892 Trinil find *Pithecanthropus "erectus"* if he didn't think that it represented the first, or one of the first, human-related species to walk upright, Leakey and Dart probably wouldn't have named OH 7's species *"Homo" habilis* if they hadn't assumed it was among the first of the hominin species to manufacture stone tools.

That assumption has been shown, however, to be wrong. As we've seen in other contexts, two cut-marked bones dated to 3.4 million years BP were found in Ethiopia's Afar Region in 2010 and were followed, five years later, by direct evidence of deliberately manufactured stone tools found in 3.3-million-year-old deposits at Lomekwi 3 on the west side of Lake Turkana in Northern Kenya.

The oldest fossil to be identified with *Homo habilis* is a 2.3-million-year-old upper jaw from Hadar in Ethiopia. Two other fragmentary finds—one from Sterkfontein in South Africa (Sts 19), the other from a 2.4-million-year-old layer at Chemeron in Kenya—lend credence to the idea that *habilis* could have been around at this time. The earliest evidence found so far of a hominin whose morphological traits "align more closely with *Homo* than with any other hominid genus" turned up in the form of a 2.75- to 2.8-million-year-old left-side jawbone, complete with its teeth, from the Ledi-Geraru research area in Ethiopia's Afar Region. This find is thought to represent a species intermediate between *Australopithecus* and *H. habilis*.

In the April 23, 1999 edition of *Nature*, Berhane Asfaw, Tim White and others announced the discovery on the Bouri peninsula on the Middle Awash River in Ethiopia, of yet another hominin living in the 2.5- to 2.6-million-year timeframe, that they named *Australopithecus garhi*. Because the two partial skulls attributed to *garhi* display several points of similarity with *A. afarensis*, Asfaw's team concluded that *garhi* "is descended from *Australopithecus afarensis* and is a candidate ancestor early *Homo*." The word "garhi" means "surprise" in the Afar language and the new hominid lives up to that name in several different ways. To begin with, it has a very small cranial capacity for such a late-appearing australopithecine: only about 450 cc. Secondly, its cheek-teeth—especially its premolars—are wide enough to blur the distinction between gracile and robust australopithecines. If *garhi*'s cheek-teeth are unexpectedly big, however, its canines are *startlingly so*—bigger, in fact, than those of any other hominin. Like the robusts (and like *afarensis*), it had a sagittal crest to anchor some of the big muscles that powered those teeth.

Post-cranial or "below the skull" bones (including partial skeletons) were found elsewhere at Bouri, in the same level as the skull or "cranial" material. If we assume that those post-cranial remains represent the same species as the partial crania, then *garhi* has a further set of surprising features. In apes and australopithecines the bones of the upper arms and the thighs are approximately the same length. In hominins with "modern" body proportions, however, such as *Homo erectus* and *H. sapiens,* the femurs or thigh bones are significantly longer than the humerus bones. *Garhi's* thigh bones are modern in this respect—that is, long in comparison with the bones of its upper arms, but that advanced characteristic stands in sharp contrast with "primitive" forearms that are as long as those of a chimpanzee—or a typical australopithecine.

* * *

Also in the Bouri area and at the same 2.5- to 2.6-million-year-old geological horizon in which Asfaw and his associates found these hominin remains, the late Jean de Heinzelin capped a long and productive career as a geologist and archaeologist as part of a team who found the left lower jaw of a medium-sized "alcelaphine" antelope—that is, a member of the grass-eating wildebeest-hartebeest family. This alcelaphine jaw had been modified in a remarkable way: on its "posteriomedial" surface—that is, on the back end of the inside of that jaw—are three successive, curved scratch/cut marks, unmistakably made by a stone tool. The position and direction of those "curvilinear striations" tell us that they were probably made in the process of removing the animal's tongue. In the same level at which this cut-marked jaw was found and about 200 yards away, de Heinzelin's team discovered a section of a large bovid's tibia that bears both cut- and chop marks. Meat had clearly been sliced off that tibia before it was smashed to extract its marrow.

Further excavation at this locality resulted in the discovery of the fairly intact femur of a hipparion or three-toed horse, bearing marks showing that it had been disarticulated (i.e., cut loose from the bones to which it had been connected) and defleshed by a stone tool.

Interestingly enough, no stone tools were found in direct association with any of these cut-marked bones or with any of the hominin remains. Stone tools were picked up at Bouri in places where the layer containing the hominin remains and the butchered bones had eroded to the surface but only in rare instances, involving one implement at a time. At Gona, however, about sixty miles from Bouri, stone tools were made and discarded in great abundance. This is so because Gona, unlike Bouri, has a plentiful supply of fine-grained stone suitable for toolmaking. The material used to make the stone tools used to butcher the antelopes and the three-toed horse at Bouri seems, in fact, to have been brought from Gona.

The only thing we can deduce for sure from these Bouri and Gona finds on the Middle Awash River is that hominins of *some* kind were using stone tools to butcher animal carcasses 2.5 million years ago. De Heinzelin and his colleagues were careful to point out that it is merely likely, but by no means certain, that the hominin identified by Asfaw, White et al. as *garhi* was that maker.

* * *

In 1984, the remarkably complete skeleton of a tall, long-limbed hominin boy dated to 1.5–1.6 million years ago (and similar to, or identical with, the *Pithecanthropus erectus* that Rene Dubois found next to the Solo River in Java) was found by Kamoya Kimeu, an associate of Meave and Richard Leakey, on a bank of the Nariokotome River near Lake Turkana in Kenya in Northern Kenya. Skull fragments attributed to the same species had been discovered at Swartkrans in South Africa in 1969. This new species—which I'll refer to as *Homo erectus*, although some paleoanthropologists regard its African form as a separate species that they would call *Homo ergaster*—wasn't just a larger version of the australopithecines. *Erectus*'s arms had become proportionately shorter than those of the australopithecines, its legs considerably longer and its shoulders, wider. Its body proportions were now, in fact, very similar to those of *Homo sapiens*. The only obvious difference between it and *sapiens* lay in the shape of their heads: the capacity of *erectus*'s brain had only increased to about 850 cc—about halfway, therefore, between the 500-cc australopithecine capacity and the 1,375 cc of *Homo sapiens*.

Erectus is often presumed to be a descendant of *habilis*, probably because the cranial capacity of specimens attributed to the latter taxon exceeds that of other australopithecines. It's not impossible, however, that more than one of the australopithecine species might have been converging on a larger-brained form toward the end of the Pliocene. We know very little about the "bush" of hominin species that existed at the start of the Pleistocene. This confusion is exemplified by our attempts to classify the famous KNM-ER 1470 skull. This 1.9-million-year-old cranium, minus teeth or lower jaw, was unearthed at Koobi Fora east of Lake Turkana (formerly Lake Rudolf) by Richard Leakey in 1972. It had a brain capacity of 775 cc—larger, therefore, than the 650 cc of the original find attributed to *Homo habilis* but smaller than the 800 cc that is regarded as the lower limit for *Homo erectus*.

Leakey didn't assign 1470 to a particular species, describing it simply as "an early member of the genus *Homo*." In 1986, a Russian paleon-tologist, Valerii Alexeev, labeled it *Pithecanthropus rudolfensis*. The *Pithecanthropus* part didn't take, but *rudolfensis* did. As a result, many people started referring to 1470 (and the handful of other finds that resemble it) as *Homo rudolfensis*. Recently, Meave Leakey and her coworkers pointed out what they see as striking similarities between 1470 and *Kenyanthropus platyops*, the 3.5-million-year-old contemporary of *Australopithecus afarensis* we talked about earlier in this chapter. They suggest, on the strength of those similarities, that 1470 be reassigned to the genus *Kenyanthropus*.

It would be entertaining if Meave Leakey's suggestion were adopted and if it were shown, thereafter, that modern humans are descended from *Kenyanthropus rudolfensis* rather than from *Homo habilis*. How, one wonders, would paleontologists rearrange their taxonomy to avoid having to call our species *Kenyanthropus sapiens*? The possibility that 1470 is close to the line of human ancestry is not, moreover, a remote one. According to Ralph Holloway, an anthropologist at Columbia University, the 1470 skull is the earliest found to date whose interior surface contains a concavity corresponding to Broca's area—a brain region that contributes to the ability of modern humans to speak. Holloway's analysis of this skull also revealed a noticeable degree of "brain lateralization." "Strong brain lateralization" is a uniquely human characteristic. It's a state of affairs in which one of the hemispheres of the brain—usually, but not always, the left one—is significantly bigger than the other.

Strong lateralization is associated with the development of the brain's so-called "language areas." Broca's area—the one that is thought to have left its imprint on the inside of 1470's skull—was discovered in the 1860s. The French physician and anatomist Pierre Paul Broca (1824–1880) performed an autopsy on the brain of a man nicknamed "Tan" or "Tan-tan." This patient had been suffering from what Broca termed "aphemia," what we now call "aphasia,"— the inability to communicate by using speech—and couldn't say anything but "tan." The autopsy showed that Tan's aphasia was associated with a syphilitic lesion of the hindmost, lower end of the frontal lobe of the brain's left hemisphere. Subsequent investigations by Broca confirmed that disruption of this particular area by accident or disease interferes with the ability to speak. Then, in 1874, Carl Wernicke (1848–1905) discovered that lesions of an area on the posterior, upper section of the temporal lobe in the same hemisphere also produce aphasia. Broca's and Wernicke's aphasias present differently: when people suffering from the former are able to speak, they do so in a halting, ungrammatical way, but they are able to make sense—and they can understand language. Those suffering from Wernicke's aphasia are unable, on the other hand, to understand what is being said to them and—even though they are often able to utter words and phrases quite freely—their speech makes no sense.

Broca's and Wernicke's areas appear, therefore, to play different but complementary roles in the business of producing and understanding speech. Connected by a thick bundle of neurons called the arcuate fasciculus, they are among the principal "islands" in an "archipelago" of language-related areas scattered across the cortex of the brain's larger hemisphere (that is, as we've seen, usually the left one). Parts of the right hemisphere also make contributions to linguistic communication: prosody—the rhythms, intonations and "melodies" that add meaning to our speech—are composed and interpreted there. The great majority of the brain's language-related functions are, however, performed in the "archipelago" which includes Broca's and Wernicke's areas. That archipelago constitutes a physically *large* system and this is at least part of the reason why the hemisphere that contains it is significantly bigger than the other one.

The inside of 1470's skull provides, as we've seen, the first direct evidence of human-like lateralization, but there's convincing *indirect* evidence that brain lateralization may already have been present at an even earlier stage of hominin evolution. That evidence concerns "handedness"—that is, the preference for using one hand over the other.

Like humans, apes prefer one hand over the other for the performance of tasks requiring a relatively precise degree of control. That preference is more or less evenly divided, however, between left and right hands in ape populations. Right- and left-handedness exists, in other words, in ape species, but no ape species is dominated (as our species is) by right-handers (or, for that matter, by left-handers). This more or less even distribution between left- and right-handedness is accompanied by a relatively weak degree of brain lateralization. In our species, on the other hand, 90 percent of whose members are right-handed, the brain is *strongly* lateralized.

If we could show, therefore, that an early hominin species consisted mainly of right-handed individuals, we'd have reason to believe that the brains of that species would already have approached the "strong lateralization" that characterizes modern humans. Surprisingly, there is a way of determining the ratio of right- to left-handers in early hominin populations. Beginning with the work of Nick Toth, several investigations have shown that stone flakes belonging to the Oldowan Industrial Complex were, in the great majority of cases, struck off their "cores" by right-handed individuals.

We're able to assume, therefore, with a fairly high degree of confidence, that the brains of those Oldowan toolmakers were already characterized by a degree of lateralization that approached that of our species. The roots of language could well extend, therefore, as Raymond Dart believed they did, well into the evolutionary history of the australopithecines.

CHAPTER 11

The Human Family as Meat Eaters and Tool Users

Hiking in the Okavango Delta in the Southern African country of Botswana some years ago, two friends and I came across a badly injured buffalo bull. The animal's stomach and testicles had been raked open, and a short length of intestine was looped over one of his horns. (That loop might, we thought, have come from the lion that had inflicted the buffalo's wounds.) It's easy to let nature take its course when you're at a decent remove, but this animal was dragging the contents of his abdominal cavity over the grass right in front of us. After a short debate on ethics, and on the rules of the Moremi Game Reserve, the ranger-guide accompanying us acceded to our request to shoot it. Like a subway passenger surprised by a lurching start, the bull stumbled sideways as the bullet hit his shoulder. He regained his footing, however, and stood there looking just as distressed as he had before. I asked the ranger if he was going to give the animal another round but was told that the ammunition was expensive, and one was going to be enough. Sure enough, the buffalo started relaxing and looking sleepy. After a minute or so, he lay down on his side. We let a decent interval go by before throwing pebbles at his eyes to make sure he was dead.

After a further debate on ethics and the rules of the Moremi Game Reserve, I took a knife out of my backpack and cut a chunk of steak from next to the animal's spine. I had to give the handle of that knife several hard punches with the heel of my hand to get its point through the buffalo's tough, sparsely haired skin, and wonder, as I write this, whether the crystalline edge of a freshly-struck Oldowan flake might not have done a better job.

Muscle fibers on the cut edges of the meat were still twisting as I wrapped it in newspaper and stowed it in my backpack. We resumed our hike and set up camp about a mile further on. It was getting dark by this time, and we made a fire, barbecuing the meat on a bed of thornwood embers. In a restaurant those steaks might have been sent back for being too tough, but their taste and flavor were so marvelous that a general realization dawned that we hadn't taken enough. As we walked back to the carcass at five o'clock the next morning to see what had become of it, wishful thinking was telling me that I might be able to get some more.

There was, however, almost literally nothing left of the buffalo. All we could see was a greasy stretch of flattened grass where it had been lying, a dark patch where its blood had soaked into the earth, and a few small fragments of bone and cartilage. The animal's head itself, with its massive horns, was gone, as was the spine and all the other large bones. The bush was dense at that spot, so some of those big body parts could have been hidden close by, but it's as likely that they'd actually been eaten. Here's what Richard Estes's *Safari Companion* tells us about the way the spotted hyena *Crocuta crocuta* deals with bones:

> Utilizes carcasses of large vertebrates more efficiently than other carnivores, which waste up to 40% of their kills. Eats everything but rumen contents and horn-bosses of the biggest antelopes.... Bones, horns, hooves, even teeth are digested completely within 24 hours.

As we saw previously, the first evidence of the first tool-assisted butchery by hominins shows up at Dikika, Ethiopia some three-quarters of a million years before the Pliocene would give way to the Pleistocene. We'll deal presently with the issue of whether those ancient meat-eaters gained access to the carcasses by scavenging or by doing the killing themselves. For now we can content ourselves with the non-controversial fact that, by 3.4 million years ago, some hominins were cutting meat off the bones of animals as large as buffaloes.

Antelopes aren't generally meat-eaters, but some duikers augment their fruit-centered vegetarian diets with the flesh of small vertebrates. The order of which humans are members, the primates, is not, in general, a meat-eating group either, but some of its members have the same kind of "anomalous" liking for meat as duikers do. These include humans, chimpanzees and baboons on the "catarrhine" or Old World side of the family, and capuchin monkeys on the "platyrrhine" or New World side. The fact that meat-eating is an exceptional and relatively recent practice in the primate order is reflected in the fact that primate body structures aren't adapted to it in any obvious way. Even though humans eat a relatively large amount of meat, they still have the comparatively broad, flat cheek teeth that were evolved to process a diet dominated by fibrous plant material, rather than the "carnassial" or meat-cutting ones possessed by true carnivores like cats.

In the 1950s and early '60s it was still widely assumed that the closest relatives of our species—gorillas and chimpanzees—were pure vegetarians. It was generally accepted that the human family must have started eating meat after it had split off from the gorilla-chimpanzee line. That "post-split" incorporation of meat into the human diet was thought at the time to have been a revolutionary and momentous innovation—one that had, in the words of a 1963 conference paper, "triggered human evolution, and propelled man to the creature he is today."

It was then discovered that chimpanzees are also meat-eaters. When Jane Goodall first reported this fact, it was thought that she'd witnessed an aberrant phenomenon—one that was peculiar, perhaps, to the Gombe Stream area in which her observations were made. It only sank in gradually that all bands of the species *Pan troglodytes* contain individuals who hunt animals and eat their flesh on a regular basis. *Troglodytes*'s close relative, *Pan paniscus*, the bonobo, still thought of as a vegetarian long after the meat-eating ways of the former became known, is now known to kill and eat flying squirrels, small forest antelopes and monkeys.

The main study group of chimps in Gombe Stream National Park in Tanzania, consisting of about forty-five individuals, is estimated to eat over 1,000 pounds of meat per year—that is, just over 20 pounds of meat per chimp. As much as half of this annual intake is consumed during the two dry-season months of August and September, a time when plant food becomes relatively scarce, and chimpanzee body weights go down.

Most, but not all, hunting is done by male chimpanzees. Chimps have larger canine teeth than early hominins had, but they seldom, if ever, use those teeth to kill their prey. They get a firm hold on the animal, rather—usually by the back legs—and flail it to death against the ground or a tree trunk. Chimpanzees who end up in control of a carcass sometimes share pieces of it with hunting participants, political allies and/or potential sexual partners. Craig Stanford of the University of Southern California reports that most prey animals are caught by cooperative action at Gombe Stream. Some 2,000 miles northwest of Tanzania's Gombe Stream, in the Ivory Coast's Tai Forest, Christophe Boesch has observed an equally high degree of cooperation in chimpanzee hunts.

To date, chimpanzees have been seen to eat the meat of between thirty and forty vertebrate species. At Gombe Stream, the young of bushbucks, duikers and bush pigs appear on their menu. When her baby son was living with her at Gombe, Jane Goodall took special precautions to protect him, because, even though it had happened a considerable time before her arrival, chimpanzees were reliably reported to have killed at least two human infants in that area for food. The Gombe chimps also kill young baboons, but colobus monkeys are by far their most commonly taken prey. Most of their mammalian prey consists of immature individuals, but some 25 percent of the colobuses they kill are adults. Craig Stanford estimates that Gombe's main study group of chimps kills an astonishing 20 percent of the colobus monkeys living in their eighteen-square-kilometer hunting range in an average year. "...[C]himpanzees," he concludes, "may be among the most important predators of certain prey species in the African ecosystems in which they live."

* * *

While the anthropological world was absorbing this "chimps are effective predators" revelation, comparisons of blood proteins were establishing another new realization: chimpanzees and humans are each other's closest relatives. (The chimp-human line had, in other words, split off from the ancestors of gorillas before it had split into its separate human and chimp branches.)

These two insights—that chimps and humans are each other's closest relatives and that both species eat meat—put the origin of human meat-eating in a new light. It's still *thinkable* that the human family "invented" meat-eating after its split from the chimp family (and that chimps had thereafter "invented" it independently of us), but the simpler and more likely explanation is that the common ancestor of humans and chimps originated meat-eating before the split, and then passed that adaptation on to its descendant species. If this is the way things happened, then the species from which chimpanzees and humans are descended would almost certainly have started its meat-eating career by hunting small animals the way chimpanzees still do, rather than as scavengers.

The earliest direct evidence of meat-eating by our family comes, as we saw in the previous chapter, from the butchered bones of medium-to-large-sized ungulates in the 2.5- to 2.6-million-year-old horizon at Bouri and Gona. The hominin remains found at different localities within that horizon at Bouri have been assigned to a single species which its discoverers named *Australopithecus garhi*. You may recall that the tool-scratched lower jaw of a "medium-sized alcelaphine antelope" was one of the cut-marked remains found at that horizon. (Alcelaphines are, as we've seen, antelopes in the hartebeest-wildebeest family.) A medium-sized alcelaphine like the hirola or "Hunter's antelope" weighs around 160 pounds. The cut-marked femur of a hipparion, or three-toed horse, found in close proximity to the jaw, represented a species that would probably have been at least as big as a hirola. The butchered "large bovid" tibia found some 200 yards from the alcelaphine jaw represents, presumably, a still-larger animal.

Back in the 1960s, stone tools had been discovered in association with the bones of even larger animals at Olduvai Gorge in Tanzania in layers just under 2 million years old. Together with rodent, tortoise, catfish and mollusk remains, 1.8-million-year-old deposits at Olduvai contain the butchered bones of not only antelopes and buffaloes, but of mammals as large as rhinos, hippos and elephants.

Raymond Dart was convinced that his little australopithecines could kill Africa's largest animals. "The animals slain by this man-ape were," he insisted, "neither all small or slow; they were huge and active." Dart's idea that australopithecine hominins hunted and ate other animals was, as I'll try to show in this chapter, correct. It's only his assumption that australopithecines hunted Africa's *biggest* animals—its elephants, hippos and rhinos—that requires qualification. When Dart made that assumption, it wasn't yet known that one of the australopithecine hominins was, by the beginning of the Pleistocene, undergoing an increase in body size that was transforming it into *Homo erectus*—a species that was, by about 1.6 million years ago, to supplant both the australopithecine species that gave rise to it and any other gracile australopithecine species that may have been around at that time. As we'll see presently, therefore, the hominins who were hunting elephant-sized animals near the beginning of the Pleistocene may no longer have been australopithecines.

* * *

By the 1980s a reaction had, as we've seen, set in against Dart's view that the australopithecines could have hunted other animals. It began in 1981 when Bob Brain's book, *The Hunters or the Hunted? An Introduction to African Cave Taphonomy*, demonstrated, as we saw in Chapter 8, that the fossilized bone fragments at Makapansgat were not (as Dart thought they were) selected and modified for use as tools and weapons. The "modifications" that had, in Dart's mind, turned them into daggers, blades, scrapers, clubs and axes were, Brain demonstrated, consistent with the effects of chewing by animals such as hyenas and porcupines.

Lewis Binford's *Ancient Bones and Modern Myths*, which appeared in the same year as Brain's book, struck another blow against the idea that early members of the human family could have hunted other animals. It suggested that the stone tools and the animal bones that had been found together at places like Olduvai might have been brought together *coincidentally* by, say, the action of flowing water. The mere fact of spatial proximity did not, in Binford's view, amount to proof that such tools had been used to butcher the bones they were found with. The proof that Binford was demanding was to emerge, however, soon after the appearance of *Ancient Bones*. Henry Bunn of the University of Wisconsin identified cut marks on bones found on the Karari escarpment in Northern Kenya's Koobi Fora district—cuts that could only have been made by hominins using stone tools to slice through the ligaments attached to those bones and carve flesh off them. Soon afterward, Pat Shipman and Richard Potts found similar marks on bones from other sites in Kenya and in Tanzania.

Did the discovery of these cut-marks turn things around and convince the anthropological community that early hominins *did*, after all, have the capability of hunting large animals? The answer to that question is "no." From about 1980 to the advent of the twenty-first century and beyond, it was, as Robin Dennell put it in a 1997 comment in *Science*, "profoundly unfashionable" to talk about big-game hunting, or indeed *any* kind of hunting, by early humans. "Nowadays," Ian Tattersall could still tell us in his 1998 *Becoming Human*, "the notion of active, early human hunting of sizable mammals has largely been discarded…"

The "early humans couldn't hunt" theory espoused by people like Tattersall and Binford didn't just deny that the australopithecine hominins could hunt—almost everybody accepted that until the 2.5-million-year-old butchered bones we talked about in the previous chapter came to light at Bouri in 1999. Binford et al. were arguing, rather, that later-developing, larger hominins like *Homo erectus, H. neanderthalensis* and, indeed, earlier members of *sapiens* itself, could not have been effective hunters.

Not everyone bought the "early humans didn't hunt" theory. "I suspect," Richard Leakey wrote in 1994, "that the recent intellectual revolution in archeology has gone too far, as often happens in science. The rejection of hunting in early *Homo* has been too assiduous."

* * *

If it's true that humans only became effective hunters very recently—that is, within the last 50,000 years, then how do paleoanthropologists explain the fact that animal bones clearly butchered by human-made tools are already present in 2.5-million-year-old deposits? Hominins got hold of such bones, the "humans didn't hunt until recently" school explained, by scavenging. The adherents of that school sold scavenging as an "objective" or "dispassionate" explanation for early hominin meat-eating. Initially, they explained, we'd resisted the idea that hominins had scavenged to feed themselves because scavenging was an "undignified" and "unromantic" pursuit. Now, however (their explanation continued), we were getting out of denial and accepting the "unflattering" notion that our ancestors had lived on the remains of hunter-predators' kills.

The idea that it's "unflattering" to be thought of as a scavenger rests, however, on the erroneous assumption that scavenging is a less *demanding* way of getting hold of meat than hunting. That misconception is related to the classic 1950s myth that scavengers like hyenas and jackals are "cowardly," "skulking" creatures, while "hunters" like lions are "noble" and "brave." There's no denying that the spotted hyena, *Crocuta*, is well-equipped for scavenging—it's able, as Richard Estes informed us at the beginning of this chapter, to chew and digest bones that are not chewable or digestible by other carnivores. *Crocuta* is, however, just as eager to get hold of fresh, meaty carcasses as lions or leopards are, and just as able to satisfy that desire by hunting. Single spotted hyenas kill fully grown wildebeest bulls, and groups of them successfully take on even larger prey.

Africa's wild savannas are divided among perpetually warring clans of these fierce, aggressive animals. Competition for food between, and within, spotted hyena clans is ferocious. Each clan patrols the boundaries of its "property" regularly, using force whenever necessary to assert its exclusive rights to occupy that territory. *Crocuta*'s most effective "enforcers" or "soldiers" are its females. "Females," Estes tells us, "not only dominate males at kills and other favored sites, but also lead clan members on pack hunts, boundary patrols and into battle."

Many of those battles are with lions. When they have a substantial numerical advantage, hyena packs can force groups of lions—especially those unaccompanied by adult males—to retreat. Male lions have been seen, in their turn, to single out, pursue and kill their enemies' big, aggressive female leaders. It's no exaggeration to say that wars of varying levels of intensity are fought on a permanent basis between groups of these two species. Because lions have the upper hand in most direct confrontations, some lion prides actually obtain *most* of the meat they eat by stealing hyenas' kills.

Scavenging is not, therefore, something that "noble" animals like lions leave to "undignified" creatures like hyenas or hominins. It is, in fact, often a more desirable option than hunting, and the privilege, accordingly, of the most powerful carnivores. "Carnivores scavenge when they can," Pat Shipman tells us, "and hunt when they must." Why should a predator expend its energy on the chancy business of chasing and killing prey when it can take an already-killed animal from a competitor who will retreat when challenged? Lions and hyenas rob leopards and cheetahs (and each other) at every opportunity. Always on the lookout for "easy meat," these two top predators pay close attention to the movements of vultures, using those birds as "eyes in the sky" to get early information about kills made in their territory.

This vigilance means that the remains of animals that meet their death in a wilderness ecology will usually be located and devoured *quickly*. It's absurd to suggest, therefore, as Ian Tattersall does in his *Becoming Human,* that the tool-marked bones found in association with some Neanderthal living areas could have been obtained from "older individuals who might have died naturally." In wilderness ecologies, "dying naturally" almost invariably means being dispatched by a predator or predators and then being consumed without delay by your killer(s) and an eager retinue of mammalian, avian and invertebrate scavengers. It simply isn't possible that bodies of animals who had died of old age would have been available frequently enough to supply hominins with a regular source of meat. There's no question that Neanderthals would occasionally have chanced upon seriously injured animals like the buffalo my friends and I came upon in the Okavango, or found helpless young deer. In the overwhelming majority of cases they would, however, have had to hunt the animals whose meat they wanted to eat, or get to a kill quickly, and drive its "owners" away with a phalanx of spears and a hail of stones before the latter had a chance to consume too much of their property.

* * *

Stealing a predator's kill before its "owner" has eaten enough of it to satisfy its own needs is referred to, for obvious reasons, as "aggressive" or "confrontational" scavenging. Aggressive scavenging can shade off into the "aggressive begging" we see when a group of spotted hyenas harasses lions who've eaten their fill, until the latter finally leave the kill to get away from the annoyance.

At the opposite end of the scavenging continuum, we find "passive scavenging." That's where the food seeker waits for the hunters and/ or aggressive scavengers to go away and then walks into the kill-site unopposed to look for whatever edible scraps they may have left. This, Lewis Binford thought, was the only kind of scavenging early humans indulged in. He acknowledged that carcasses utilized in this way would not usually have any meat left on them. "The major, or in many cases, the only usable or edible parts consisted," Binford wrote, "of bone marrow."

The hominin-butchered bone-fragments discovered at Bouri (and discussed in the previous chapter) confirm that early hominins did break open bones to get at their marrow, but the cut-marks found on many, if not most, of the Bouri bones or bone-fragments prove that the bones in question were defleshed *before* they were smashed open. The position of the cuts on one of them shows, in addition, that it was "disarticulated"—that is, cut off from a larger fragment or from a whole carcass.

The only kind of "nonconfrontational" scavenging that might have allowed our species regular access to flesh-covered carcasses, is (as

I see it) the theft of the gazelle-sized carcasses that leopards often store in trees. Given the fact that australopithecines were agile tree climbers, I imagine that "larcenous" scavenging of this kind must have taken place relatively often. That possibility does not, however, solve the mystery of how hominins living in the Pliocene and early Pleistocene came into possession of the flesh-covered limbs of mammals larger than gazelles. The medium-sized alcelaphine, the three-toed horse, and the "large bovid" whose remains were butchered by Bouri hominins are outside the size range of animals normally killed by leopards, and the human-butchered buffaloes, hippos and elephants found at the early-Pleistocene sites we'll discuss later are, of course, clearly so.

Shortly after the Bouri bones were discovered, further sets of cut-marked bones were discovered in 2.6- to 2.3-million-year-old layers about sixty miles away at a locality called Gona. Manuel Domínguez-Rodrigo et al. spelled out some of the implications of these finds in the following abstract of a contribution to a Paleoanthropological Society conference published in April 2003:

> Newly discovered archaeological sites at Gona (Ethiopia) preserve both stone and faunal remains. These sites have also yielded the largest sample of cut-marked bones known from the time interval 2.6- to 2.3 million years ago (Ma). Most of the cut-marks on the Gona fauna possess obvious macroscopic (e.g., deep, V-shaped cross-sections) and microscopic (e.g., internal microstriations, Herzinian cones, shoulder effects) features, that allow us to identify them confidently as instances of tool-imparted damage caused by hominin butchery. In addition, the anatomical placement of the cut-marks on several of the recovered bone specimens suggests that Gona hominins eviscerated carcasses, and defleshed fully muscled upper and intermediate limb bones of ungulates—activity which further suggest hominins gained early access to large animal carcasses. These observations support the hypothesis that the earliest stone artifacts functioned primarily as butchery tools and also imply that

hunting and/or aggressive scavenging of large ungulate carcasses may have been part of the behavioral repertoire of hominins by c. 2.6–2.5 Ma, although a larger sample of cut-marked bone specimens is necessary to support the latter inference.

The cautious note on which this abstract closes is restricted, of course, to the 2.6- to 2.3-million-year-ago time frame—we've accumulated so much evidence of hominin butchery of large animals of the late Pliocene and early Pleistocene, and accumulated it, moreover, in so many different parts of Africa, that such butchery can no longer be regarded as a fluky or aberrant occurrence. In addition to the Pliocene sites we've just been talking about, cut-marked bones have been found at early-Pleistocene sites including Kenya's Koobi Fora region, Olduvai Gorge in Tanzania, and South Africa's Sterkfontein Valley. The cumulative message sent to us by these finds is that the hominins of that time were somehow getting access—on a fairly regular basis—to animal carcasses that had flesh on their bones. As I've already argued, it doesn't ring true that those hominins would, in the normal course of events, have gained possession of flesh-covered bones or carcasses by passive scavenging. Regular access to valuable resources of that kind could, in my view, only have been obtained by hunting or by confronting other meat-eaters.

When we consider the "other meat-eaters" that existed in late-Pliocene, early-Pleistocene Africa, confrontational scavenging begins, however, to look like an even more unlikely alternative than hunting. In addition to lions, leopards and cheetahs, the Africa of that time was inhabited by three species of "machairodont" cat: the scimitar-tooth *Homotherium*, the "classic" sabertooth *Megantereon*, and the dirk-tooth *Dinofelis*. Early Pleistocene Africa was, as we've seen, also inhabited by the enormous *Pachycrocuta* hyena.

It's difficult to see how the australopithecine hominins who butchered the Bouri and Gona bones could have defended their kills against—let alone stolen meat from—a fearsome animal like *Pachycrocuta*. This giant hyena lived throughout Africa up until about 1.4

million years ago (and stayed in existence a good half-million years longer outside that continent). It occupied an immense stretch of territory between Spain, South Africa and China, and occurred, as one might expect from such a widely spread genus, in a variety of forms. Paleontologists distinguish *P. pyrenaica, P. bellax, P. robusta* and *P. brevirostris. Pachycrocuta's* largest forms were as big as a present-day lion, and the structure of its teeth and jaws shows that it was, like its close relative the still-existing spotted hyena *Crocuta,* specially adapted to the business of crushing and eating bones.

Just as we were tempted, therefore, in the later twentieth century, to say "early hominins must have scavenged because they couldn't possibly have hunted," reflection on the formidable size and power of *Pachycrocuta,* and on the big cats that existed along with it, could drive us to the opposite conclusion: that the hominins of 2.5 million years ago *must* have obtained the carcasses they butchered by hunting, because they couldn't possibly have scavenged! I'll argue presently that the late-Pliocene australopithecines who processed the cut-marked bones found at Bouri and Gona did obtain many or most of the carcasses they butchered by hunting. We can surmise that the hominin hunters of 2.6 million years ago might only have been able to defend those carcasses against relatively manageable challengers such as cursorial, striped and brown hyenas, surrendering them, perhaps, only to more frightening ones like *Pachycrocuta* hyenas and big cats. Gaining the ability to drive off even *some* kinds of predator-scavengers would have meant, however, that they had also gained the ability to steal meat from those species. It seems reasonable to assume, therefore, that late-Pliocene australopithecines of the kind that butchered the bones found at Bouri and Gona, might have augmented their hunting activities with both larcenous and confrontational scavenging.

* * *

Adjectives like "crude" and "simple" aren't used as frequently as they used to be to describe Oldowan cutting flakes, since we've become aware of how much knowledge and skill was required to make those tools. Their makers would have had to know, firstly, how to identify the "isotropic" rocks required for their manufacture—that is, those that split in any direction, rather than along a single preferred plane like slate. Pliocene toolmakers must, therefore, have been able to identify high-quality tool-making materials like obsidian, flint and chert, and passable ones like fine-grained quartzite. They would, it seems, assay the qualities of likely-looking rocks in the same way a modern geologist does—by hitting them with a hammer (except of course that their hammers would have been stones). "You could tell," relates Craig Feibel, a geologist-anthropologist who investigated the 2.3-million-year-old Lokalalei 2C tool-making site west of Lake Turkana, "that they would pick up a cobble that was coarse-grained, hit it a couple of times, and see it wasn't good. Then they would pick up a fine-grained rock, see it was the right quality, and keep chipping, and knock thirty flakes off a single core."

A considerable degree of technical skill would have been required to produce those flakes. By teaching himself to reproduce tools of the Oldowan pattern, Nick Toth has found out that their manufacture involves a great deal more than just banging two rocks together. You have to find a "shelf" forming an "acute" or less-than-90-degree angle in the rock you've selected for flaking—or make one—and then deliver an accurate blow about half an inch from its edge with your "hammerstone," making sure that the force of your blow travels through the right area of the "core" rock to detach a usable flake.

The late-Pliocene hunter-foragers we've been talking about didn't sit around at a home base making the tools they would take with them on tomorrow's or next week's hunting expedition. Later hominins often did things that way, but Oldowan tools were typically made at the butchery site where they were used. We know this because flakes produced in the course of the manufacturing process are liberally scattered around many kill sites, and because "manuports"—raw material in the form of non-local rocks—are also sometimes found at such sites. We know it, too, because archaeologists have repeatedly managed to reassemble the original cobbles from which flakes were struck at the places where those tools were used.

The fact that the butchers of the late Pliocene made their "meat cutters" at the places where they used them shouldn't lead us to believe, however, that they would only have sought out and collected the raw materials needed to make those cutters *after* a carcass had been secured. Many or most butchery sites would have been situated miles away from outcrops of workable rock. Gona is, as I've mentioned, the nearest source of tool-making rock to Bouri where the carcass-processing we've been talking about was done. If the hominins who gained control of the Bouri carcasses had to walk the sixty miles to Gona to fetch tool-making rocks *after* they'd secured the carcasses in question and then walk sixty miles back, the meat would, of course, have been long gone by the time they got back. It's generally accepted, therefore, that the Oldowan toolmakers of the late Pliocene and early Pleistocene would, in many cases, have taken their tool-making cobbles with them on their hunting-foraging expeditions, carrying them as unopened packages of fresh blades, so to speak, in anticipation of securing a carcass that would require butchery.

How did those toolmakers carry the cobbles they took with them on those expeditions? Chimpanzees have been seen to carry the rock "hammers" they use for opening certain kinds of nuts for a little over 500 meters, but the hominin toolmakers of the Pliocene carried their "unopened packages of blades" for many miles. It's highly unlikely that they would have carried those rocks in their hands. One of their hands would have been employed, as I'll argue presently, in carrying a spear, club or stick. My own experience of walking in a wild African savanna tells me that the other hand would have been in constant use, steadying their passage over a rocky outcrop here, and gingerly pushing a thorny branch out of their way there. It would also have been used to pick the berries and other fruits they would have been passing, to shade their eyes, to feel the temperature of fresh dung, and to chase off biting insects. That "free" hand might have been used for signaling, too, at times when it was critical to maintain silence, and to work with the other hand in the manipulation of weapons.

It's just not realistic, in short, to assume that the early-Pleistocene hunter-foragers we've been talking about would have carried their tool-making rocks in their hands. Could they have suspended them from their shoulders or necks in animal-skin slings? Hominids may well have started using "bags" of that kind (not, of course, created by sewing of any kind) for gathering vegetable foods by the relevant time, but hunting and aggressive scavenging would have involved trotting, running at full tilt, and executing all the demanding body movements that would have been part of the business of bluff, attack, evasion and retreat. A heavy, loose object hanging from the neck or shoulder would interfere with those kinds of movements. It seems possible to me, therefore, that the Oldowan toolmakers may have bound their rock-carrying "bags" snugly against their bodies in a rough, folded or rolled kind of "fanny pack." Tying the simple knots required for this kind of device need not, in my view, have been beyond the capabilities of a hominin who'd acquired the demanding skills required to create stone cutting-tools.

* * *

Now for the matter of weapons. I've argued that the hominins who butchered the cut-marked bones found at Gona, Bouri and elsewhere must have obtained the majority of those bones either by hunting or by confrontational scavenging. Since late-Pliocene hominins weren't nearly as big or as powerful as Africa's largest predators, and since they weren't equipped with the kind of teeth or claws that those carnivores possess, they could not have hunted animals as big as, or bigger than, themselves (or have confronted the big meat-eaters that would have been competing with them) without hand-held weapons like wooden staffs, clubs, spears or "distance weapons" like thrown rocks.

Their spears would *not* have been tipped by stone points. The development of composite weapons of that kind lay more than a million years into the future. Their spear-points were made by simply sharpening wooden shafts. Oldowan cutting tools weren't only used for butchery: they were also used for working wood. A 1981 *Nature* article by Lawrence Keeley and Nicholas Toth reports that the authors established the existence of microscopically distinct wear polishes on products of the so-called Karari Industry, a more recent variant of the Oldowan Industrial Complex (the implements in question date from about 1.5 million years ago). Nine of the fifty-four implements examined showed wear traces. On three of those nine, the traces were of the same kind as those the authors produced experimentally by working wood with stone tools they had made themselves. One of the three wood-marked tools displayed the kind of crescent-shaped damage scars that the authors produced experimentally by using stone flakes as saws.

It seems likely, therefore, that stone cutting-tools might have been used from their inception to shape wooden implements like digging sticks and spears, and to sharpen their wooden points. Hominids may, in fact, have been sharpening sticks and bones by rubbing them against abrasive rocks *before* the advent of stone cutting tools. Adrienne Zihlmann has underlined the importance that the gathering, transportation and sharing of plant foods must have had for the development of hominin technology. It's likely, in this view of things, that digging sticks were used before the advent of manufactured stone tools. The technological distance between a digging stick and a spear is not a long one.

I'll have more to say about the development of tools and weapons presently. Let's step back first however, and take a look at the hominins who processed the 2.5-million-year-old cut-marked bones that we've been talking about. If they're *Homo habilis*, or something like that species, then the tops of their heads are probably round, like those of chimps or humans. If they're *Australopithecus garhi*, then humps of muscle and fat on top of their heads might have given them a gorilla-like look. (Some of *garhi*'s big chewing muscles were, like those of gorillas, anchored to a bony crest running across the top of its skull.) The bodies of such hominins probably look more human-like than their heads, but their forearms remain disconcertingly long and ape-like. *Garhi*'s thighs are, however, probably longer than those of a typical australopithecine. *Garhi* is, for this reason, taller than the other hominins of that time, but still considerably shorter than modern humans.

If we could travel back in time and habituate a group of late-Pliocene hominins to ourselves the way Dian Fossey and Jane Goodall did with gorilla and chimp groups, we'd probably come to regard some of them with the affection and respect Fossey felt for Digit, and Goodall had for Flo. We'd like and respect them, though, as fellow creatures rather than fellow humans. The sexual and emotional parts of our minds would probably insist that beings like *Australopithecus garhi* and *"Homo" habilis* are vastly different from ourselves.

Imagine "alien" species such as these walking upright like humans through the late-Pliocene/early-Pleistocene bushveld. About a million years will go by before the family of which they are members will learn to use fire, but these beings do not exist in a state of technological naivety. They could be carrying digging, fruit-hooking and hunting sticks, and one or more of them might, as I've suggested, be wearing rolled skins containing raw materials for the manufacture of stone tools and, perhaps, a favorite hammerstone. Those rolled "fanny packs" might also contain non-utilitarian but "interesting" objects like shells or stones that caught the eye of their finder. A remarkable jasperite manuport from Makapansgat in South Africa, brought to that site by a hominin between 2.5 and 3 million years ago, has two holes representing eyes and another representing a mouth. These features weren't carved into that flat, oval stone—the stone's resemblance to a hominin face is accidental but probably the reason why it was noticed and picked up.

Thanks to Bernhard Zipfel, Curator of Collections and the Evolutionary Studies Institute, University of the Witwatersrand, Johannesburg, for permission to use this image.

The argument that stones of various kinds were carried in skin pouches is, to my mind, a strong one. If pouches were in fact used in this way, one could speculate that they might also have been used to transport food and even water stored in naturally formed containers like those grown by the Cucurbitaceae or gourd family. Infants could, by this time, also have been tied against, or suspended from, their mothers' bodies in animal-skin holders.

It's central to Darwin and Wallace's theory of evolution by natural selection that the bodies and behaviors of living beings change relatively gradually, over thousands of generations. The fact that such change takes place at a *slow* rate does not mean, however, that it takes place at a *constant* rate. It probably proceeds, on the contrary, by what Darwin himself described as "slow and interrupted steps."

Let's consider, for a moment, the "slow and interrupted steps" that might have led from the killing and eating of squirrels and infant monkeys by our ancestors, to the hunting and butchering of large antelopes. It's probable, as we've seen, that those ancestors started hunting before they started walking upright. At this stage they would, like present-day chimps, have been killing their prey without hunting weapons, "butchering" it—that is, tearing it apart—with their hands and their teeth.

Chimps can, as we've seen, catch, kill and dismember animals the size of a twenty-five-pound colobus monkey with their bare hands. It's quite possible, therefore, that hominins may also have been able to kill and butcher animals of this size without using weapons or tools. It's less likely, but still possible, that they could have used their bare hands to catch, kill and dismember animals as big as the fifty- to seventy-pound ground-dwelling *Paracolobus* monkeys that lived in the late Pliocene and early Pleistocene. It begins to look highly unlikely, though, that they could have killed and dismembered 100-pound animals like, say, gerenuk antelopes in this way. By the time hominins were eating the meat of antelopes weighing 150 pounds or more, we know for sure (from the archaeological evidence we've been discussing in this and the previous chapter) that they were butchering those animals with stone blades and scrapers. If they gained control of many of those big antelopes by either killing them or stealing them from their killers—an unavoidable conclusion in my view—then it's a practical certainty that they used weapons for both those purposes.

The development of weapons was, in Charles Darwin's opinion, intimately linked with our species' bipedality. "The hands and arms," he wrote in his *Descent of Man*, "could hardly have become perfect enough to have manufactured weapons, or to have hurled stones and spears with true aim, as long as they were habitually used for supporting the whole weight of the body..."

"Darwin was arguing," Richard Leakey comments on this passage, "that the evolution of our unusual mode of locomotion was directly linked to the manufacture of stone weapons." Since manufactured stone implements only came into use about 2.5 million years ago (we now know they were made and used almost a million years earlier) and humans have been walking on two legs since at least 6 million years ago, Leakey reasoned that the two phenomena could not have been related. "Whatever the evolutionary force that produced the bipedal ape," Leakey concluded, "it was not linked with the ability to make and use tools."

It's true that Darwin linked the development of bipedality with the manufacture of weapons, but he didn't, as Leakey suggested he did, link it to the manufacture of *stone* weapons. The modification of stone to make tools or weapons must (as Darwin himself pointed out in *The Descent*) have been preceded by a long period during which unmodified stones were used for those purposes. Contemporary observations that chimpanzees sometimes throw stones at competitors and predators, and that some chimps use stones to break open nuts, underscores the reasonableness—and, in fact, the *obviousness*—of this commonsense assumption.

In a very real sense chimpanzees already "manufacture" tools—they pluck, strip and trim the twigs they use for termite "fishing," prepare the branches that they use to hook fruit-bearing branches into reach, process the leaf masses that they use to sponge water out of hollow trees by chewing them, and sharpen crude spears with their teeth to skewer bush babies hiding in hollow trees.

The same logic that tells us that the chimp and human families probably inherited their meat-eating ways from their common ancestor suggests that the two families derived their tool-using abilities from the same source. It's by no means unlikely, therefore, that hominins were using a variety of objects as tools and/ or weapons (in both modified and unmodified states) before they started walking upright.

Darwin's guess that the development of weapons and upright walking could have stimulated each other's development remains, therefore, in my view, a good one.

Darwin believed, in addition, that the positive feedback loop that linked upright walking and the use of implements included the development of our species' extraordinary level of intelligence. Richard Dawkins questions that linkage. "...[T]his is a rare instance," he argues, "where Darwin's tentative guess has turned out wrong. The fossils give a satisfyingly clear and decisive answer. Bipedality came first, and its evolution was more or less complete before the brain started to swell."

The fact that the early-Pleistocene butchers of the Middle Awash had small brains need not, however, invalidate Darwin's hypothesis that the development of tool-use and bipedalism were linked to an increase in hominin intelligence. Dawkins's argument against it rests on the assumption that there's a more or less linear correlation between a primate's brain size and its intelligence: that is, if australopithecines had chimpanzee-sized brains, then they must have had chimpanzee-sized mental capacities. It's clear, though, that the australopithecine butchers of Gona and Bouri used their chimpanzee-sized brains to gain control of the body parts of large ungulates, and disarticulate and deflesh those parts with cutting implements. These achievements are very far beyond the intellectual abilities of chimpanzees. The intelligence of at least one hominin species had, therefore, already exceeded that of chimpanzees to a very significant extent while the size of its brain was still close to that of a chimp. The superior intelligence that characterizes our species may well have gotten its start, therefore—as Darwin suspected it did—by stimulating, and being stimulated by, our species' use of tools and weapons.

CHAPTER 12

Bodies, Brains, and Technology

THOMAS EDISON COULDN'T HAVE PRODUCED HIS LIGHT BULB if a host of previous inventors hadn't made discoveries ranging from the manufacture of glass to the production of electricity. Edison didn't even come up with the basic idea of passing an electric current through a resistant filament inside an oxygen-free glass bulb: a patent based on that concept had already been granted to Matthew Evans and Henry Woodward in 1875, and several other people had, by that time, produced light bulbs of one kind or another. Edison's contribution to the production of a commercially viable light bulb took the form of buying Evans and Woodward's patent and tweaking it—mainly by looking for a filament that would burn longer than those currently in use. When he started that search in 1878, a light bulb developed by Sir Joseph Wilson Swan, an English physicist, had just set a record of 13.5 hours. By examining literally thousands of plant species, Edison found a cotton-derived fiber that delivered 40 hours of light in 1879 and then, in 1880, a Japanese bamboo fiber that extended that time to 1,200 hours. The bamboo fiber was replaced by a tungsten filament a few years later, but by that time, Edison's "invention of the light bulb" had, in the general opinion, already become a fait accompli.

We mustn't conclude from this account that Edison's work on light bulbs was overrated. All inventors rely more or less heavily on the work of their fellow inventors, and Edison's investigation of the things that happen inside light bulbs was to leave a particularly large number of inventors indebted to *him*: he had learned that electricity could flow across the vacuum inside those bulbs, from the hot filament to a positively charged metal plate. (He'd introduced the metal plate in an unsuccessful attempt to draw off the black deposit forming on the inside of a light bulb.) The discovery of this "thermionic emission" or "Edison effect" led directly to the invention—by one of Edison's engineers—of the vacuum tube. That device was to be of crucial importance in the development of radio, television and computers, and would eventually stimulate a search for the semiconductor that would perform its functions in more powerful and versatile ways.

It's in the nature of technological advance to breed fresh technological advances. Our family's entry into the "cognitive niche" we'll speak about in Chapter 15 did not, therefore, simply lift hominins up onto a new plateau of ecological power and then leave them there. It started, instead, to propel their technological development along a growth curve that rose slowly for a long time before climbing abruptly into the ever-steepening ascent that it's presently making.

* * *

I suggested, in Chapters 10 and 11, that an enhanced ability to make tools and weapons 2.5 million years ago may have allowed at least one of the australopithecine species to hunt medium-sized herbivores and to stand its ground against at least some of the carnivores that would have been competing with it for that kind of prey, or for access to their carcasses. For a hominin population developing these kinds of abilities, group action with wooden spears and thrown rocks might eventually have replaced tree-climbing as the prime defensive strategy against its competitors and predators. That kind of shift in strategy would presumably have relaxed—and perhaps reversed—the selective pressure that had, up to that time, been restricting that and other hominin species to a size that would permit fast and agile tree-climbing.

For this and/or other reasons *Homo erectus* had, by 1.7 or 1.6 million years ago, become, in Richard Leakey's words, "tall, athletic and powerfully muscled." "Even the strongest modern professional wrestler," Leakey tells us, "would have been a poor match for the average *Homo erectus.*" A million years later, the bodies of *H. heidelbergensis*, a European descendant of *erectus*, were still exceptionally big and strong. The Schöningen javelins we'll talk about presently are heavy enough to suggest that they were, as Robin Dennell put it, "best used by large, powerfully built people. The extremely robust Boxgrove tibia appears to have belonged to such a person."

The female members of *erectus*-to-be underwent an even bigger growth spurt than the males, diminishing the big size difference between the sexes that had characterized the australopithecines.

* * *

The dramatic and relatively abrupt increase in the body size of the species that was becoming *Homo erectus* was accompanied by an equally radical change in the shape of this new hominin's body. Its forearms became shorter, while its legs, and more especially its thighs, grew longer. Its narrow shoulders gave way to broad ones, while its waist grew long and relatively slender. Beginning with an article by Denis Bramble and David Carrier in a 1983 issue of *Science*, titled "Running and Breathing in Mammals," a number of theorists have argued that these changes in body shape came about in response to the demands of running.

How, one might ask, could running have played a part in our evolution if it didn't offer our species some kind of advantage over other animals? Don't humans run much more *slowly* than most other animals? The answer is that, slow as they are, humans are able to outrun a great many other species over *long distances*. We're able to do this because we've evolved highly effective ways of dissipating the heat generated by sustained physical activity. Natural selection has elaborated and multiplied our "eccrine" sweat glands to the point where it's quite inappropriate to speak of "sweating like a pig": humans perspire far more heavily than pigs, and, indeed, most other mammals.

The cooling caused by the evaporation of our sweat is enhanced by the fact that our skins are largely hairless. The selective pressure that stripped us of our body hair must have been powerful because there would have been significant countervailing demands for the retention of that covering. Much of the territory in which hominins evolved is high country in which sub-zero temperatures are common on winter nights. Just as hairlessness could, and probably did in some cases, cause hominins to freeze to death on winter nights in Ethiopia's mountainous regions and on South Africa's highveld where snowfalls are not uncommon, it would have exposed them to sunburn on hot days and contributed, under such conditions, to the dehydration and demineralization risks posed by the prodigious power of their eccrine sweat glands.

When did hairlessness emerge? Newborn human infants still exhibit a reflex that produces a grip strong enough to bear their own weight. This "gripping reflex" likely evolved to grasp the mother's body hair. The fact that it still exists suggests that hairlessness must have evolved relatively recently. It seems reasonable to assume, therefore, that it might have come into being in concert with the other cooling mechanisms that contribute to the extraordinary long-distance running capabilities that *Homo erectus* seems to have possessed by some 1.7 million years ago.

Because of a variety of specialized adaptations, such as kidneys that minimize water loss, some desert antelopes can go without drinking water indefinitely. Oryxes, dik-diks and others with adaptations of this kind, let their body temperatures rise to levels that would be lethal to their brain tissue rather than lose moisture by using water to cool their core temperatures by sweating. They're able to do this by selectively cooling, through the evaporation of mucus inside their nasal cavities, of the blood that flows to their brains. The cooling system that *Homo erectus* evolved is, by contrast, designed to cool its entire body: its core temperature. Could the evolution of a cooling system that makes such prodigal use of water have been facilitated, one wonders, by the fact that the hominins in question had learned to transport and/or cache water in calabash shells, the shells of ostrich eggs, or other "ready-made" containers of that kind?

Having "opted" for full-body cooling, our genus does not have the elaborate vascular networks that many dry-country antelopes have evolved for the selective cooling of their brains. It does, nevertheless, seem to be capable of some degree of differential brain cooling. In a 2004 article in *Nature*, Denis Bramble and Daniel Lieberman argue that our venous circulation may be "designed" to carry blood cooled by sweating on the scalp and face, back into the interior of the skull to cool, by countercurrent heat exchange with arterial blood on its way to the brain by way of the internal carotid artery. Dean Falk argues, too, that emissary foramina—tunnels that conduct brain-cooling veins from the scalp through the skull to the dura mater or outer cover of the brain—were significantly increased in the gracile australopithecines that split off from their robust counterparts between 3 and 2.5 million years ago.

* * *

The 2004 *Nature* article by Bramble and Lieberman that I've just quoted, is the most comprehensive argument to date, that the anatomy and the physiology of *Homo erectus* were extensively adapted to long-distance or endurance running. The demands created by walking are, these authors point out, very different from those created by running. Swing your right leg forward while walking, and you will, Newton's third law of motion tells us, create an equal and opposite reaction that will tend to turn the rest of your body on its axis in a clockwise direction. That potentially destabilizing reaction is counteracted by our hip muscles: via the left foot, which is in contact with the ground, they exert a countervailing, anticlockwise torque that—quite literally—keeps us on track.

Running creates larger rotational forces than walking. Those forces cannot, however, be counteracted through a connection with the ground because running bodies become airborne between toe-off and heel strike. Natural selection has, therefore, elaborated a different way of counteracting that rotation: as the pelvis rotates in a clockwise direction, to sweep the left leg forward, the chest makes a sharp anticlockwise rotation to neutralize the reaction of the legs and pelvis. "Sawing" arms attached to wide shoulders add momentum to this countervailing torque.

The chest of a running hominin like *Homo erectus* could not counter-rotate as effectively as it did if its connections to the rest of the body—that is, to the head and the lower body—were not highly flexible. This is the reason, Bramble and Lieberman argue, why *erectus* evolved a waist that was narrower, longer and more flexible than that of a chimpanzee or an australopithecine. We must, however, qualify this "long and flexible waist" argument: Dr. Guillaume du Toit, a spinal surgeon practicing in Cape Town, informs me that much of the torsion that allows our species to rotate its upper body and shoulders so effectively takes place in a region of the thoracic spine near the level of the shoulder blades.

The need to counter-rotate the chest and shoulders is also, Bramble and Lieberman argue, the reason why *erectus* developed the long, flexible neck that still characterizes our genus. In chimpanzees, the head is connected to a pair of narrow, "shrugged" shoulders by numerous powerful muscles, advantageous for tree-climbing, and this, these authors assert, was also the case with the australopithecines. In *Homo erectus* and its descendants, those muscles are reduced in size or completely absent. The result is a head that can maintain an "even keel" during running, rather than being made to "yaw"—rotate back and forth on its axis—by the rotation of the shoulders.

Bucking this trend toward reducing the muscular connections between the head and the trunk, *erectus*-to-be evolved a so-called "nuchal ligament"—a connection between the back of the head and the spine which had the effect of preventing its head from pitching back and forth in response to the heel-strikes and toe-offs of running. Ligaments of this kind are, Bramble and Lieberman argue, present in species that are either adapted to running, or massive. The skulls of *Australopithecus*, they report, lack the "nuchal shelf" to which these ligaments are attached.

* * *

Increasing the ability of the chest and shoulders to rotate independently of the lower body and the head didn't only allow *erectus*-to-be to *run* more efficiently—it also permitted it to become an excellent *thrower*. Bramble and Lieberman argue that this "decoupling" of the chest from the pelvis and the head is what allows our stance to remain firm, and our heads to retain a fixed orientation to the target, while our chests and arms rotate rapidly and forcefully with the throwing motion.

The Schöningen javelins we'll talk about presently provide direct evidence that *Homo erectus*-like beings were using throwing skills to hunt by nearly 400,000 years ago. We've seen, though, in Chapters 9 and 10, that one or more hominin species was hunting medium-sized prey by 2.6 million years ago. It seems likely, therefore—especially in view of the fact that *Homo erectus*'s distinctive body shape is already fully developed in the partial skeleton of the

1.6-million-year-old "Turkana boy" discovered by Richard Leakey in 1984—that a selective pressure in favor of forceful, accurate throwing had already started to operate in the late Pliocene.

I suspect, for these reasons, that when we see a baseball- or cricket player execute a sizzling, flat-trajectory throw from the outfield straight into the hands of a catcher or a wicketkeeper to run someone out, we're witnessing a skill that was already in existence in the early Pleistocene. And if the big, powerful hominins of the early and middle Pleistocene were able to throw rocks with that kind of velocity and accuracy, then carnivores—even big carnivores like machairodont cats and *Pachycrocuta* hyenas—would have risked serious injury by remaining in throwing range of a group of them.

* * *

The ability to throw missiles hard and accurately would have given its possessors obvious advantages in hunting and in conflicts with other predators who were trying to seize them and/or their prey. Exactly what advantage did endurance running confer on the genus *Homo*?

In "The Energetic Paradox of Human Running and Hominin Evolution," David Carrier argued the thinkable: that *Homo erectus*'s ability to run in a "perseverant" way had evolved in response to that species' strategy of pursuing swift, open-country prey. (A capacity for endurance running could not, after all, help a hominin to escape from a predator.) Apparently unaware of Carrier's writings on this subject, Wilhelm Schüle reached the same conclusion independently in the 1990 "Paleo-ecology of Hunting" article we discussed earlier. "Tropical carnivores," Schüle wrote,

> attack with great speed and succeed instantly or otherwise abandon their hunt. Most of them hunt by night. They cannot tolerate chasing prey for any amount of time in tropical heat.

Man, the naked, sweating animal can do just that. Any well-trained jogger can outdistance his dog on a hot summer day. The dog's weight/surface index is more favorable than man's, yet its cooling system is far less efficient.

Endurance running obviously isn't needed to pursue the kind of prey that chimpanzees hunt: tree-dwelling monkeys, young forest antelopes and squirrels. It would, however, be needed, as we've just seen, to hunt animals like savanna antelopes and members of the horse-zebra family. These are precisely the kinds of animals whose carcasses were butchered 2.5 million years ago by the late australopithecine hominins we discussed in Chapters 9 and 10.

Given the logical fit between the evidence that suggests that those 2.5-million-year-old antelopes and equids may have been hunted by the hominins who butchered them and Bramble and Lieberman's "our bodies were shaped by long-distance running" thesis, it's surprising to learn that these two authors still adhered to the orthodoxy that hominins did not hunt until recently. How, given that view, do they explain their contention that endurance running was important enough to our late-Pliocene ancestors to exert a profound influence on the evolution of our anatomy and physiology? Our need for endurance running could, they suggest, have been created by scavenging. "Wild dogs and hyenas," Bramble and Lieberman argue,

> often rely upon remote olfactory and visual cues such as circling vultures to identify scavenging opportunities, and then run long distances to secure them. Early *Homo* may have needed to run long distances to compete with other scavengers...

I have to say—with respect for an article which is, in the main, well-reasoned—that this is totally unconvincing. The distances over which kills can be seen by humans and, indeed, seen or smelled by their four-legged competitors aren't nearly long enough to give endurance runners like ourselves an advantage over other animals. Bramble and Lieberman themselves admit that the "well-conditioned human runners" who can "occasionally outrun horses" only manage to do so over "extremely long distances."

Circling vultures aren't visible to humans at "extremely" or even moderately long distances. Investigations by the Northern Prairie Wildlife Research Center showed that a black-painted model vulture with a seven-foot, ten-inch wingspan, was "barely visible" to an observer on the ground when released from an aircraft at 4,700 feet (i.e., about 0.9 of a mile) above the ground, and invisible without binoculars at 5,800 feet (about 1.1 miles). Most of Africa's larger mammals would be able to outrun our species with ease over a one- or two-mile distance. The African wild dog, *Lycaon pictus*, a hunter that specializes in running down medium-sized antelopes can, according to Richard Estes's *Safari Companion*, run at thirty-five miles per hour for at least two miles, while elite human middle-distance runners can barely manage fifteen miles per hour over that distance. Even hyenas would easily outdistance the fastest human over a two-mile distance.

I argued in Chapter 10 that *erectus*-to-be was probably engaged in both larcenous and aggressive scavenging by the beginning of the Pleistocene, but it seems clear that the ability of this species to maintain a slow running speed over very long distances was evolving in response to the other strategy it employed to gain access to animal protein, namely hunting.

* * *

Perhaps because Bramble and Lieberman reject the idea that recent human groups might have employed endurance running as a hunting strategy, they conclude, after a perfunctory discussion of this issue, that endurance running is "not common among modern hunter-gatherers." That statement may be correct with regard to the hunter-agriculturalists who live in rainforests—an environment that our species has only entered in relatively recent times—but there's clear evidence that many, if not most, of the hunter-gatherer groups who were living in open country during the nineteenth and twentieth centuries did use endurance running to secure their prey.

Bramble and Lieberman themselves quote Peter Nabokov's *Indian Running: Native American History and Tradition* in a footnote, but they do not discuss Nabokov's accounts of numerous Native American tribes using endurance running to secure a wide variety of swift-running prey ranging from jackrabbits to deer.

There are many other accounts of North American Indians tiring out their prey by perseverant running. A classic book by Wendell Bennett and Robert Zingg, *The Tarahumara: An Indian Tribe of Northern Mexico*, published in 1935, describes how members of that group chased down deer until the animals were exhausted and then throttled them to death by hand. In his *Why We Run: A Natural History*, Berndt Heinrich tells us that the Paiutes and Navajos were reputed to do the same with pronghorn antelopes. Heinrich reports, too, that Australian Aborigines chased down kangaroos by forcing them to reach lethal body temperatures.

It has also been frequently and reliably documented that, well into the twentieth century, the San hunter-gatherers of Southern Africa's Kalahari region killed antelopes and other herbivores by pursuing them at an aerobic pace for many hours. San hunters could and did run down unwounded animals, but, even in cases where their quarry had been hit by a poisoned arrow, they often needed to run after it for long distances to prevent it from being eaten by a competing carnivore or recovering from the effects of the poison and making its escape.

* * *

The genetic evolution that was giving *erectus*-to-be its formidable long-distance running abilities and its relatively big brain was, in the early Pleistocene, still "tightly coupled" (as E. O. Wilson put it in his *Consilience*) with the cultural evolution that was producing its toolkit. Genetic and cultural evolution weren't only moving at a more or less similar pace—they were, in addition, probably linked to each other in a positive feedback loop: the development of more sophisticated tools might have contributed to the selection of brighter tool-users, who might then have produced still more sophisticated tools and so on.

Today these two kinds of evolution are no longer walking hand in hand. Technological evolution is moving at such enormous speed that it has completely outpaced the "traditional" or gene-based evolution that made it possible in the first place. Nobody would suggest, for instance, that the physical and mental capacities of the human species have undergone any noticeable change since the beginning of the nineteenth century, but the technological progress that has, since that relatively recent time, taken us (as an article in *Fortune* magazine put it) "from steam engines to search engines" has been so phenomenal that it challenges our ability to comprehend it fully. The minds and bodies of present-day humans probably aren't even significantly different from those of the people who created the sophisticated images of Europe's big animals in Lascaux Cave some 17,000 years ago, but the development that our species' technology has undergone since that time has, from the Lascaux artists' point of view, been completely unthinkable. Cultural evolution has raced ahead of its gene-based counterpart in this way, because the latter is limited, as we've seen, to piecing together useful new structures or behaviors from randomly occurring genetic variations over many generations, while our family's ability to innovate ontologically—to invent—can literally bring useful new behaviors into existence overnight.

But the faculty that *produces* ontological innovations could not, of course, have arisen overnight. The neural machinery that runs inventive intelligence was, on the contrary, assembled—by gene-based evolution—over many thousands of generations. Regardless, therefore, of how explosive the growth of our family's technological power was eventually to become, its initial development would not have been fast. The archaeological record confirms that this was the case: the Oldowan toolkit produced by *garhi* and/or its contemporaries in the late Pliocene was very simple in comparison with later human artifacts. What little change it would undergo over the next million years would, moreover, manifest itself very slowly.

Slow change is not, however, the same as no change. "Over the several hundred thousand years evident in Olduvai's stratigraphy," Roy Larick and Russel Ciochon tell us, "Oldowan assemblages undergo distinct refinement in chipping techniques and some standardization in tool form. By 1.7 to 1.6 mya, bifacial tools help to define the Developed Oldowan Industry."

In the region of 1.6–1.5 million years ago, those bifacial tools developed into the assemblage of heavy-duty scrapers, cleavers, picks and "hand axes" characteristic of the so-called "Acheulean" or "Mode 2" industry. Acheulean tools were dramatically bigger than their Oldowan counterparts, and more efficient: "I can work much faster with a cleaver or a hand axe than I can with a small flake," Nick Toth reports. "I can make longer sweeps with them, I can cut deeper into the meat, and my fingers don't get as tired."

Usually about as big as an open human hand, the teardrop-shaped "hand axes" were used as knives and saws rather than axes. The "cleavers" with their wide, chisel-like front ends were probably used, inter alia, for the separation of the skin and muscle, while the Acheulean "picks" could have been used for the same purposes as modern picks are—to loosen up earth for digging purposes.

Although the Acheulean industry is named after Saint-Acheul in France where its products were first found, it only appears in the European archaeological record in the F layer of Notarchirico in Southern Italy, a horizon established by thermoluminescence dating to be about 640,000 years old. By 500,000 years ago, it is, however, found throughout Europe, from Boxgrove in England to Korolevo in Ukraine.

* * *

The surge of technological development that led to the invention of these new stone tools in Africa between 1.6 and 1.5 million years ago may have included a more varied and effective range of wooden implements. I suggested in Chapter 11 that wooden implements like digging tools and spears must already have been used by the makers of Oldowan stone tools. We saw, too, in that chapter, that Lawrence Keeley and Nicholas Toth found wear polishes confirming that some of the "developed" or late-Oldowan stone tools they examined were used to cut or saw wood.

Archaeologists have tracked hominin woodworking into the early Acheulean period. Manuel Domínguez-Rodrigo and his associates reported in the 2001 volume 40 of the *Journal of Human Evolution*, pp. 289-299, that wear patterns on the edges of 1.6-million-year-old hand axes from Peninj in Tanzania and phytoliths adhering to those edges, reveal that those tools were, in fact, used as "knives" and "saws" to cut a variety of substances including wood. We know, therefore, that woodworking of some sort was taking place in the early days of the Acheulean industry but have not, to date, found any of the wooden implements that would have been produced at that time. That's not unexpected: unless wood happens, by some odd chance, to end up in an anoxic environment such as a peat deposit, it decomposes relatively quickly.

A few wooden implements have been found in association with *later* Acheulean technology. Apart from wooden artifacts produced near the end of the African Acheulean around 200,000 years ago at Kalambo Falls in Zambia, all these finds have been made in Europe. There's an irony in that state of affairs. Africa was the "big city" of hominin activity during the Acheulean period, while Europe was a distant suburb that only received that technology long after it was developed in Africa. Now that Europe has become a "big city" of industry, activities like civil engineering, mining, building and farming (not to mention archaeology) are carried on there much more intensively than they are in Africa. This means that more Acheulean material gets unearthed in formerly "backward" Europe than on the continent on which that industry was developed.

The most spectacular collection of Acheulean-produced wooden implements discovered to date came to light in the 1990s near the town of Schöningen in Lower Saxony, in what Robin Dennell describes as "one of the many enormous, unsightly, opencast brown-coal mines that dot the German landscape." Destructive as this mine may have been of the present-day German landscape, its excavation revealed a succession of long-buried middle-Pleistocene environments. One of the levels it uncovered, number 4 of the Schöningen II channel, included a lakeshore. The pollen, mollusk shells and other organic fragments recovered from this layer represent life-forms that were alive between 380,000 and 400,000 years ago, during an episode of relative warmth just after the fifth-last interglacial maximum, which is known in Germany as the Holstein interglacial. Examining a cache established by human, or hominin, hunters near the shore of that ancient lake, Hartmut Thieme and his associates found the remains of a hearth in which those hunters had made fires, together with hundreds of stone tools, and a great many flakes produced in the process of retouching those tools. A large number of butchered animal remains, mainly those of horses, were also unearthed there, together with a finely-worked thirty-inch length of spruce wood, almost an inch and a quarter in thickness, and sharpened to points at both ends.

Remarkable as the latter find was, it was overshadowed by the discovery of seven well-preserved wooden spears. These spears, which range from six feet to eight feet four inches in length, are between 1.2 and 2 inches in diameter at their thickest points. Each was made from the whole trunk of a young, approximately thirty-year-old, spruce tree. Those young trunks had been debarked and their *Astansätze*—the beginnings of their side branches—painstakingly worked down. The spears' points were carved from the bottom ends of the trunks, where the wood is at its densest. Apart from making for hard points, this means that the heaviest, thickest parts of the spears are, like those of modern javelins, situated toward the front. Those heavy front ends were—also like those of present-day javelins—tapered down gradually into slender, sharp points, leaving the spears' center of gravity about a third of the way from the front.

There had been previous indications that the pre-*sapiens* inhabitants of Europe used wooden spears to hunt big game. In 1948 in Lehringen, Germany, a reasonably well-preserved yew spear was found inside the remains of an elephant dating from the last interglacial 125,000 years ago; in 1911 the tip of what could have been a spear was found at Clacton-on-Sea in England in deposits laid down at more or less the same time as those in which the Schöningen spears were found; and a round hole in a rhinoceros scapula found at Boxgrove in England, dated to the sixth interglacial before the most recent one—that is, about 500,000 years ago—provides possible evidence of a spear wound.

The world of Anglo-American paleoanthropology is, as we saw in Chapter 10, still dominated by the idea that humans only started hunting big animals near the end of the Pleistocene, around 40,000 years ago. "To fit this picture," Dennel explains,

> the Clacton and Lehringen spears were downgraded to digging sticks or, imaginatively, snow-probes for locating buried carcasses.

> But the Schöningen discoveries are unambiguously spears: to regard them as snow-probes or digging-sticks is like claiming that power drills are paperweights.

Even before the discovery of the Schöningen spears, it had been widely accepted among German anthropologists that the early European members of the genus *Homo* were expert big-game hunters. Perhaps the body of archaeological evidence in support of that proposition that has been found on the continent of Europe is simply too physically big to deny. Fully 60 percent of the enormously abundant bone fragments that hominins had processed and accumulated at Bilzingsleben in Thuringia are, for instance, those of big animals like the "forest" or "straight-tusked" elephant *Paleoloxodon*, woodland rhinos, aurochs, horses and bears. The remaining 40 percent are evenly divided between middle-sized animals like red and roe deer, and smaller vertebrates like beavers, fish and birds. The Bilzingsleben site was occupied at a time or times lying between 410,000 and 380,000 years ago.

Large, pre-Neanderthal accumulations of animal remains, with megafauna also represented heavily, have been found in association with hominin tools at several other places in Europe, including Fontana Ranuccio and Isernia La Pineta in Italy. Remains of large animals dated between 300,000 and 700,000 years BP have been found in association with stone tools in many European localities, including Toralba and Ambrona in Spain, Arago and Terra Amato in France, and Notarchirico in Italy. An excavation made in the process of constructing the undersea rail link between England and France uncovered a 400,000-year-old, partially butchered skeleton of an elephant surrounded by stone tools at Ebbsfleet in Kent. Butchered elephant carcasses dating from between 300,000 and 500,000 years BP have also been found just above the B layer of Notarchirico and at Aridos I and II in Spain.

Whether we refer to it in general terms as *Homo erectus "senso latu"* ("in the broad sense") or speak more specifically of it as *H. heidelbergensis*, it's clear that the hominin species living in Europe 500,000 years ago was an expert big-game hunter. It was a hunter that seems, moreover, to have enjoyed ascendancy over Europe's other predators. Mark Roberts, a British archaeologist who shares the "Continental European" view that Europe's early hominin inhabitants were regular and successful hunters of big game, directed the excavation and analysis of the remains of three adult *Stephanorhinus* or "woodland" rhinos that were killed and butchered by *erectus* at Boxgrove in southern England, about 500,000 years ago. Each of those kills would, Roberts tells us, have been "a magnet for other predators."

> Yet each carcass was skillfully cut up. Fillet steaks were sliced from the spine and the bones were smashed to get out the marrow. Only hunters in complete control of their patch could have done that.

Where they are present on the bones of these rhinos, the tooth marks of those "other predators" overlie the marks of this butchery, showing that animals like the lions, hyenas and wolves that inhabited England at that time, only got access to those bones after the humans discarded them.

* * *

The wealth of archaeological material we've been discussing makes it clear that European hominins were already capable of killing big game some 700,000 years BP—that is, even before the Acheulean industry entered that continent. We also know, from evidence found at Terra Amata in Southern France as well as at Bilzingsleben and Schöningen, that the hominins living in those places were, by at least 400,000 years ago, already controlling and using fire.

Did these two ancient European skills—big-game hunting and fire use—arise in Europe, or were they, like the Acheulean tool-making industry, brought to that continent after being developed somewhere else at some even more ancient time? It would hardly be surprising if the latter turned out to be the case: the "big city" of human technological innovation lay, as we've seen, in the continent of Africa at this time, where the Acheulean industry had been in existence for almost a million years before it spread into Europe.

* * *

Let's talk about big-game hunting first. Are there any indications that hominins were hunting animals as big as elephants, rhinos and hippos in Africa before they were doing so in Europe?

As we saw in Chapter 9, the oldest cut-marked bones found to date, the 2.6-million-year-old specimens unearthed at Bouri, show that the tongue was removed from a medium-sized antelope, and that the haunch was cut off the carcass of a three-toed horse. Bone fragments found at nearby Gona, in a 2.6- to 2.3-million-year-old deposit, display cut-marks which suggest to Manuel Domínguez-Rodrigo and his colleagues that the Gona butchers "eviscerated carcasses, and defleshed fully muscled upper and intermediate limb bones of ungulates." It seems, therefore, that the butchers of both Bouri and Gona had early access to intact or nearly intact carcasses. Carcasses in that condition are, as I argued in Chapter 11, much more likely to have come into the possession of their butchers through hunting and/or aggressive scavenging than through passive scavenging.

I also argued, in Chapter 11, that hunters could seldom have retained control of the carcasses of larger prey animals if they were power-less over the carnivores that surrounded them. There's nothing far-fetched about the idea of *erectus* (or *erectus*-to-be) becoming able to drive off or kill big cats with spears. Many Africans killed lions with spears during the twentieth century, and Alexander ("Sasha") Siemel (1880–1970), a Latvian who emigrated to Brazil during the First World War, became famous for killing jaguars with a spear. In doing so he was, however, simply emulating a feat of a native who taught him to become a "tigrero"—that is, a hunter who kills jaguars armed only with a spear. In the article titled "Inter-viewing the Tiger Man," Siemel states: "...I learned the art from a poor native who had nothing but a homemade spear where I had my high-powered rifle. But I do think I was a good pupil and will admit that it calls for experience and judgment."

Evidence that early-Pleistocene hominins were exploiting the carcasses of Africa's biggest animals is provided by a number of "slaughter-sites" discovered in East Africa, at which the whole or nearly whole carcasses of megaherbivores were dominated by hominins for periods of time long enough to carry out more or less extensive butchery.

At Koobi Fora in northern Kenya, parts of a hippopotamus carcass, dated to 1.9 million years BP, were found in association with simple stone tools. The skeleton of a *Deinotherium,* some of whose bones were butchered and disarticulated, was discovered in Olduvai's 1.8-million-year-old FLK North Lower Bed II in Tanzania, in association with thirty-nine stone implements.

Another proboscidean—a member of the genus *Elephas,* associated with 172 stone tools—was found at FLK North, Upper Bed I, level six. This site is dated at between 1.7 and 1.5 million years BP. Olduvai deposits of this age have also produced several tool-butchered carcasses of the long-horned buffalo *Pelorovis.*

The thoroughly butchered skeleton of another *Elephas*—dated at between 1.6 and 1.3 million years BP—was found with 569 Oldowan implements at Barogali in Djibouti, a small country sandwiched between Somalia and Ethiopia. The cranial roof of this animal had been smashed away from its calvarium, presumably to get to the brain.

Is this catalog of early-Pleistocene African slaughter-sites a short or a long one? Because he was doing fieldwork in Zimbabwe during a time in which thousands of elephants were being culled in that country, the American archaeologist Gary Haynes was given the opportunity of observing firsthand the "taphonomy" or after-death fate of the remains of these animals. He was also present in areas of Zimbabwe where thousands of elephants had been shot by ivory hunters during the second half of the nineteenth century. His observations in these killing zones brought him to the conclusion that the carcasses of large animals are only preserved in highly exceptional circumstances.

The passage of 1.5 million years or more since the early Pleistocene would have further decreased the already-unlikely prospect of finding a preserved elephant carcass. When we consider in addition that Africa is, archaeologically speaking, a comparatively unexplored continent by contrast with Europe and North America, the fact that six or more butchered carcasses of megaherbivores have been found in that continent's early-Pleistocene deposits cannot be taken as an indication that hominin butcheries of such animals must, at that time, have been infrequent occurrences. It could, in fact, suggest the contrary.

* * *

We've seen that *erectus* had, by 1.6 million years ago, become larger and more powerful than present-day humans; that it could probably outrun the swiftest antelopes over very long distances; and, that it could probably deliver "distance weapons" like thrown rocks or spears with accuracy and power. These changes help to explain how *erectus* could have become an effective hunter by that time—even a very effective hunter—but they do not, by themselves, explain how that big new hominin species could have become destructive enough to drive some of its prey animals and competitors into extinction.

It's only when we return to a consideration of our family's ability to innovate on an ontogenetic level, and focus on the degree of power that that unique ability had conferred on *erectus* by the early

Pleistocene, that it can begin to seem understandable—and perhaps even inevitable—that this species brought about the hemorrhage of extinctions that depleted Africa's large-mammal diversity around 1.4 million years ago. The unprecedented degree of power that this ability to innovate ontologically was conferring on *erectus* by the early Pleistocene was, as we'll see in the next chapter, most compellingly manifested the control it achieved, during that period, over fire.

CHAPTER 13

Fire

IN 2012 ARCHAEOLOGISTS PAUL GOLDBERG, FRANCESCO BERNA and others reported the discovery of convincing evidence of hominin fire use securely dated to 1 million years BP, at the Wonderwerk Cave in the hilly grassveld of northwestern South Africa. That evidence consisted of carbonized grass, leaf and twig fragments, together with bone fragments. Many of these items had been heated to temperatures reaching 700°C. That was a crucial piece of information, because grass or brush fires typically burn cooler than 250°C, while hearths or campfires can reach and exceed the 600°C mark.

The bone fragments and the plant ash had been burned where they had been found—"in situ" in archaeological language—and had not, therefore, been washed or blown into the cave from elsewhere. A dating technique known as "cosmic burial" established that the geological stratum in which they had been found was laid down approximately 1 million years ago. Sandwiched inside that stratum, one on top of another, were the products of many other small fires. Also associated with the ash found in that stratum were numerous "pot-lid fractures"—small stone chips shaped like frisbees or pot lids that are ejected from larger stones under the influence of high temperatures. Some of those lids could be

refitted to the rocks that had ejected them, adding confirmation of the undisturbed condition of the site.

* * *

Strong evidence of fire use found in Swartkrans Cave, some 340
miles northeast of the Wonderwerk site, appears to be some
600,000 years older than the Wonderwerk material.

Swartkrans, situated in the Sterkfontein Valley about thirty
miles northwest of Johannesburg, is the single richest hominin
site in Africa. Its oldest deposits—Members 1 to 3—have yielded,
between them, the skeletal remains of two hominin species that
were living in the early Pleistocene, as well as tools made of both
bone and stone, and animal bones butchered by hominin meat-
eaters. Excavating grid square W3/S2 in Member 3 of this trea-
sure house of early-Pleistocene hominin life on March 21, 1984,
George Moenda, working with Bob Brain, found an apparently
charred piece of bone. Had this specimen come from a later
archaeological excavation, Brain tells us,

> I would have had no hesitation in assuming that it had been
> partly charred in a fire, but at an estimated age in excess of a
> million years, the presence of burnt bone in a hominid living
> area required further verification.

Brain accordingly took a sample of it to the National Physical
Laboratory of South Africa's Council for Scientific and Industrial
Research, where analysis confirmed that the blackening of the
bone was, in fact, caused by fire rather than chemical staining.

By the time the excavation of Member 3 was completed, 270
pieces of burned bone had come to light. Although several of them
had been found grouped in clusters, it's clear that those burned
bones were, for the most part, the products of a great many separate fires:

In one grid square, W3/S3, burnt bones occur in 23 excavation spits, each 10 cm deep, indicating that the bones were heated in frequently recurring fires during the deposition period of this stratigraphic unit, which may have spanned several thousand years.

The distribution of these burned bones throughout such a large portion of the six-meter-deep gully of calcified debris that constitutes Member 3 doesn't only suggest a long history of fire use in or near the entrance to the Member 3 gully; it also excludes the possibility that the burned bones could have "contaminated" that Member—that is, found their way into it at some relatively recent time.

In order to find out whether the bones were burned in natural fires, or in fires made and/or tended by the hominins who lived in that area, it was necessary to establish the temperatures to which they had been subjected. As we've seen, grass or brush fires typically burn cooler than 250°C, while hearths or campfires can reach or exceed the 600°C mark. In a test "hearth" fire made with wood from *Celtis africana*, an African relative of the elm that commonly grows in the entrances of karst caves like Swartkrans, Brain measured temperatures of up to 860°C. A test fire made with the wood of *Acacia karoo*, another tree that is presently common in the vicinity of Swartkrans, peaked at 688°C.

There is, of course, no absolute rule that brush or veld fires will not exceed 250°C. Grass fires can ignite dead trees, or parts of trees, to produce, in localized patches, the high temperatures characteristic of hearths. It's very unlikely, however, that anything beyond a small percentage of a collection of naturally burned bones (which would be more or less evenly distributed on the veld and not concentrated in the vicinity of logs) would be heated to those temperatures.

To begin the task of determining the temperatures at which the Swartkrans bones were burned, Brain experimentally heated a large selection of fresh and weathered bones in a kiln at selected temperatures. He found that the color of these bones changed to dark brown or black in the vicinity of 300°C. At higher temperatures this discoloration faded and the bones started becoming "calcined." At about 600°C they would become grayish-white in color, and tend to be lighter in weight.

In order to create a histological or microscopic record of these changes, Brain cut the fresh, defleshed radius of a hartebeest into seven transverse segments and heated those segments, respectively, to 200°C, 300°C, 400°C, 500°C, 600°C, 700°C, and 800°C for thirty minutes, cooling them slowly thereafter. Thin sections prepared from each of these segments displayed more or less unique microscopic features, characteristic of the particular temperature to which it had been subjected. Another portion of each of those bone segments was sent to Andrew Sillen, then of the Archeometry Laboratory of the Department of Archaeology at the University of Cape Town, for chemical analysis.

With the knowledge they had gained from this controlled heating, Brain and Sillen then subjected the burned bones found at Swartkrans's Member 3 to histological and chemical analyses, establishing that well over half of them had been subjected to temperatures of 300°C or more. Forty-seven percent had become calcined and grayish-white in color and were found to have been heated to temperatures approaching or exceeding the 600°C mark. Dr. Anne Skinner and her colleagues at Williams College in Williamstown, Massachusetts, have subsequently confirmed these findings by using ESR to test samples of the bones in question for the presence of the different kinds of free radicals that are created by burning at different temperatures.

The picture that has emerged from these analyses makes it highly unlikely that anything but a small minority of those bones could have been burned in natural fires. We know, in addition, that hominins ate meat off at least some of the burned bones found in Member 3 because four of them display cut- and/or chop marks made by stone tools. (No cut-marked bones have been found outside Member 3.) Of the 59,218 unburned bone fragments found in that Member, a total of twelve (0.02 percent) are cut- or chop-marked; of the 270 burned bones found there, four (1.48 percent) are marked in that way. The fact that only 1 in 4,934.8 of the unburned bones is cut-marked, while 1 in 67.5 of the burned bones is marked in that way, would be a puzzling one if we assumed a "natural" or non-hominin cause for the bone-burning, but it makes complete sense if we accept that the burning of those bones was, like their cut-marking, carried out by hominins.

With the exception of the lower bank of Member 1, in which three specimens were found, no other burned bones were found in any of the other Swartkrans Members, although each of those other Members yielded plenty of other evidence of hominin remains and tools. Could this mean that the entrance to Member 3 had at the relevant time some feature like, say, a dolomite overhang, which made it convenient to build fires in its immediate vicinity, a feature that the other Members lacked? Or could it indicate that the hominins only learned to use fire around the time that Member 3 started accumulating? Or that the hominins in question had only been capable of harvesting naturally ignited fire before the accumulation of Member 3 (which could explain why only three burned-bone specimens were found in Member 1) and that they had learned, by the time Member 3 was accumulating, to *make* fire rather than relying on the relatively infrequent occurrence of naturally ignited fires in that area? The latter two possibilities are both reconcilable with East African evidence that suggests, as we'll see presently, that fire use by hominins started between 1.7 and 1.6 million years ago, while faunal and ESR dating suggest that Swartkrans's three older Members accumulated in the vicinity of 1.6 million years ago.

Eight hundred seventy-seven stone specimens adjudged to be hominin artifacts were found in Members 1, 2 and 3. These hominin-fractured stones, which consist mostly of cores and unmodified waste, aren't easy to classify. Clark Howell, of UC Berkeley's Laboratory for Human Evolutionary Studies, thinks they "approximate the Developed Oldowan of eastern Africa..." Howell points out, however, that a number of bifaces, a pick-like piece, and a cleaver-ended specimen, obtained from dump breccia residues presumably originating from Member 3, raise the possibility that this Member contained an Acheulean element.

Member 3 also yielded forty-one bone specimens—mainly fragments of long bones and horn-cores—whose points are scratched and rounded in ways that show that they were used for digging. A total of eighty-four such implements were found in Members 1, 2 and 3, and a few more have been unearthed at nearby Sterkfontein and Driemolen. Brain concluded from experimental work done with modern replicas of those tools that the originals were used to unearth plant storage organs like the bulbs of *Ledebouria* and corms of the genus *Hypoxis*. (Both the beautiful yellow flowers of *Hypoxis hemerocallidea* and the cheetah-spotted leaves of *Ledebouria socialis* can still be seen in the vicinity of Swartkrans.) Experimental work by Lucinda Backwell and Francesco d'Errico suggests, however, that the width and orientation of the striations on those tools are more likely to have been produced by digging in termite mounds.

A delicate, awl-like artifact from Member 3, SKX 37052, has a point that is marked by both circumferential and longitudinal scratches, together with polish. "This tool may well have been used," Bob Brain and Pat Shipman wrote,

> ...for piercing holes in skins and other soft materials, as similar microscopic wear has been documented on experimentally made and used awls. The evidence discussed here suggests that the Swartkrans hominids of Member 1–3 times may well have made simple carrying bags from animal skins, in which they transported their tools, as well as possibly their gathered food. This could explain the evidence for the apparent use of the same tools over successive days or weeks.

Hominin remains—mostly fragmented like the majority of the bone specimens found in that cave—are extremely abundant at Swartkrans. Skeletal parts representing an estimated 116 individuals of the species *Paranthropus robustus* have been found in the cave's three "old" Members. Skeletal remains of *Homo*—presumably *Homo erectus*—are present at Swartkrans, too, but in much smaller numbers: fragments (including one lower jaw) represent perhaps three individuals in Member 1 and two in Member 2. The fact that no remains clearly identified with *Homo* were found in Member 3 does not mean that this genus was absent from Swartkrans area during the time that Member was accumulating: lions and sabertooth cats appear even less frequently in the Swartkrans deposits than *Homo* does, but it would be unrealistic to conclude from that fact, that big cats must necessarily have been absent, or even rare, in that area during the times in which their remains don't show up.

Which of the two human-like beings whose remains have been found at Swartkrans made the tools, and tended the fires in which the Member 3 bones were burned?

When Louis Leakey's family unearthed a specimen of both *Paranthropus* and the gracile hominin OH 7 in 1.8-million-year-old deposits at Olduvai in which stone tools had been found, they were confronted—at least in regard to the tools—with a similar question. Leakey conceded that it was possible that both hominin species had made and used those tools, but inclined to the view that the species which he called *Homo habilis* "was the more advanced tool-maker, and that the *Zinjanthropus* [i.e., *Paranthropus*] skull represents an intruder (or a victim) on the *Homo habilis* living site."

It's by no means impossible that *Paranthropus* could, on occasion, have become a victim of *Homo*. *Homo erectus* had, as we've seen, become an expert hunter of large animals in the early Pleistocene, and there's no reason to think that large primates would have been excluded from their list of prey species. *Erectus* is thought to have butchered giant gelada baboons later in the Pleistocene, on the basis of 700,000-year-old evidence found at the DE/89 B site at Olorgesailie in Kenya, and we know that contemporary humans kill and eat chimpanzees. It's interesting, too, to note in this regard that the burned bones found in Swartkrans's Member 3 include a *Paranthropus* finger-bone. (Those bones also include one each of a baboon, a zebra, a guinea fowl, a dassie and a meerkat, while the rest are those of a group of antelopes and other split-hoof animals which zoologists refer to as bovids.)

The relationship between *Homo erectus* and *Paranthropus* would have been a strange one from the point of view of present-day humans. *Erectus*—less intelligent, we can assume, than *sapiens* was to become—would have been sharing a habitat with a fellow primate that was probably quite a bit more intelligent than present-day chimps or gorillas. Randall Susman's anatomical investigations have led him to the conclusion that *Paranthropus*'s hands were well-adapted to precision grasping and the use of tools. It seems clear that *Paranthropus* and *Homo* were genetically isolated from each other, but there's no reason why simple technological memes could not have flowed between them. It would not be surprising, at any rate, if digging tools like those found at Swartkrans, as well as wooden implements of various kinds, were used by both hominins. One could speculate that *Paranthropus* might have used crude stone tools—and perhaps even the discarded Acheulean tools they would occasionally have come across. Acheulean tools are, however, so frequently and reliably associated with *Homo erectus*, and associated with it, moreover, for so long after *Paranthropus* becomes extinct around 0.9 million years ago, that there can be little doubt that *erectus* was the developer and practitioner of the Acheulean industry (and, indeed, its immediate predecessor, the Developed Oldowan). In the absence of evidence suggesting otherwise, I'll continue to presume, therefore, that advanced technologies like the making of Acheulean tools and the use of fire were the exclusive preserves of the big new hominin whose brain capacity was considerably larger than that of both its australopithecine ancestors and its robust contemporaries.

* * *

When did the fire use evidenced in Swartkrans take place? Unlike many of the bones and hominin-associated objects from East Africa, which are often conveniently sandwiched between layers of volcanic ash from which relatively reliable dates can be obtained, those found in South Africa's limestone caves are notoriously difficult to date. The fossil- and artifact-bearing deposits of Swartkrans entered that cave at different times, falling into it or sliding down "talus" slopes through different entrances or holes in its roof. At times when the cave was accessible, some objects are thought to have been carried into it by four-footed carnivores or hominins. The body of debris presently found in that cave constitutes, therefore, what Brain understatedly calls "a depositional mosaic of considerable complexity."

In a professional involvement with the cave that has lasted more than half a century, Brain has resolved that mosaic into five more-or-less distinct "Members." Those Members represent, in Brain's view, depositional episodes that alternated with erosional phases, in cycles that were probably driven by precipitation changes triggered, in their turn, by the 40,000-year glacial cycles that the planet was experiencing up to a million years ago. The deposits are numbered in the order in which they appear to have accumulated in the cave, with Member 1 being the oldest, Member 2 being the second oldest, and so on.

In the 1960s and '70s Elizabeth Vrba—then of the Transvaal Museum, later of Yale—made a detailed and comprehensive study of the fossil bovids from Swartkrans and other hominin-bearing cave deposits. On the basis of when those bovid species had made first and last appearances in deposits elsewhere in Africa—deposits that *could* be securely dated by stratigraphic and other means— she concluded in 1982 that a portion of Member 1, the so-called "Hanging Remnant," was between 1.8 and 1.5 million years old. With the exception of the burned and/or cut-marked bones we've been discussing, the hominin-associated objects found in Members 2 and 3 don't differ significantly from those of Member 1. Apart from the fact that *Homo* doesn't appear in Member 3, the hominin remains recovered from the three "old" Members don't differ from each other either. The animal ensembles found in those three Members are also substantially similar. Members 1, 2 and 3 are presumed, for these reasons, to have accumulated in relatively rapid succession to each other.

A 2003 review of the fauna of Swartkrans's Members 1–3 by Darryl de Ruiter, now of Texas A&M, yielded an estimate for the age of those Members that was very close to the result produced by Elizabeth Vrba's study of Member 1's "Hanging Remnant." De Ruiter concluded that, although Members 2 and 3 were contaminated by some recent material, all three of the older Members were deposited in the neighborhood of 1.6 million years ago.

Direct ESR testing of two hominin teeth and two bovid teeth by Rainer Grün, Darryl Curnoe and others have produced findings that are in agreement with those of Vrba and de Ruiter, yielding an age of between 1.47 million and 1.79 million years with a median of 1.63 million years for the "Hanging Remnant" of Member 1 in which those teeth were found.

As one might expect from deposits that accumulated before the 1.4-million-year-ago "hemorrhage" of African megafaunal diversity we talked about in Chapter 4, Swartkrans's Members 1, 2 and 3 contain several species that disappeared in that hemorrhage: the cheetah-like *Chasmaporthetes* hyena makes two appearances, along with the sabertooth cats *Dinofelis* (one appearance in Member 1) and *Megantereon* (one appearance in Member 3).

In my view, Bob Brain's meticulously documented excavation and analysis of the burned bones in Swartkrans's Member 3 do not merely suggest that hominins *may* have been using fire by 1.5 million years ago. They raise, instead, an overwhelming likelihood that they were doing so.

* * *

Because signs of fire use are not, until the later Pleistocene, found as frequently as, say, stone tools are, Ian Tattersall suspects that fire use may have been a "rather intermittent" part of human life in the earlier parts of that epoch. It doesn't seem likely to me, however, that fire use could have been abandoned or lost throughout the entire African and South Asian range that our family occupied in the early Pleistocene. The fact that fires feed on organic material—mainly wood—which is preserved far less frequently than stone artifacts, is, in my view, a more likely explanation for the fact that we don't find the signs of fire use as consistently as, say, stone tools.

But stone artifacts are themselves able to provide information about early hominin fire use. When "siliceous" stone such as basalt or quartz (as distinct, say, from "calcareous" material such as limestone) undergoes prolonged exposure to temperatures exceeding 250°C (which marks, as we've seen, the borderline between the "campfire" and "brush fire" temperature ranges) its surface undergoes characteristic color and texture changes, developing, in particular, dimple-like scars called "pot-lid fractures," discussed earlier. A survey by Brian Ludwig of Rutgers University of stone artifacts—and of the waste pieces or "debitage" produced in the making of such artifacts—revealed a telling pattern of thermal alteration. Ludwig personally examined almost 40,000 pieces of hominin-modified stone from nearly fifty sites in the Olduvai and Turkana basins, covering a period of 2.5 million to less than 1 million years ago. In doing so, he found no pot-lid fractures on any specimens dating from before about 1.6 million years ago. After that time, however, a small but consistent proportion of the stone does display pot-lid fractures, over a wide variety of sites including Olduvai where, despite the preservation of a rich store of hominin-modified material, no signs of fire use had previously been found.

The other area surveyed by Ludwig, the Turkana Basin, had, however, yielded signs of hominin fire use prior to his survey. These were first identified in the 1970s, when Jack Harris, who had been Ludwig's teacher at Rutgers, examined several patches of apparently baked reddish earth at a cluster of sites known as FxJi-20, situated in the 1.6-million-year-old Okote tuff at Koobi Fora. Harris and some of his colleagues became interested in those patches when they realized that they resembled the present-day spots on which local tribespeople had made overnight campfires. These apparently baked patches were situated in an area that had yielded a number of hominin artifacts, some of which were apparently discolored and/or fractured by heat. A 1978 suggestion by Harris that the reddish patches were remnants of fireplaces or hearths, provoked a response that was initially skeptical, but now tends toward acceptance. Ralph Rowlett, an anthropologist from the University of Missouri, then established, by thermoluminescence tests of four of the discolored patches conducted in association with Charles Peters, that those patches had been heated more recently than the volcanic soil in which they were situated. That considerably diminished the possibility that they could have resulted from non-thermal processes. Rowlett also eliminated the possibility that they had been directly caused by lightning strikes by examining known strike sites in Africa and the United States. He concluded that the discolored Koobi Fora patches—which were between twelve and twenty inches in diameter—would have been much smaller if they had represented lightning strikes. Those patches were not, moreover, associated with fulgurites, the glassy bits of fused sand that are produced by such strikes.

Rowlett, Randy Bellomo and others found melted crystals in the patches which indicated that the fires causing them had produced temperatures close to 400°C. We've seen that grass- or brush fires on Africa's savannas can and do ignite logs and/or dead tree stumps, which can burn hotter than the fire that ignited them. The likelihood that discolored patches had been produced in this way was tested by two different methods.

Firstly, an archaeomagnetic analysis of several possible fireplaces at Koobi Fora's FxJi-20 main site by Randy Bellomo yielded evidence that they had been reheated several times over a period of several years. It's not unlikely that campfires would be relighted repeatedly on particular spots which tradition and convenience might have established as hearths, but highly unlikely that "natural" ignition of logs and stumps would have occurred repeatedly on the same places.

Secondly, Ralph Rowlett (together with Robert Graber and Michael Davis) examined the phytoliths left by various kinds of fires, including those that had discolored the suspected Koobi Fora hearths. Phytoliths are microscopic bodies of opaline silica formed in epidermal and other cells of the growing plant, from silica dissolved in groundwater originally absorbed by the plant's roots. They typically survive both the decay and incineration of the plant that produced them, and can often be used to identify the species or, at least, the general botanical category of that plant. Rowlett and his colleagues reasoned that a fire consisting of a naturally ignited log or stump would only leave one kind, or relatively *few* kinds, of phytolith. Fires that were, on the other hand, started by hominins would have left many kinds, because they would have been kindled and fed with grass, leaves, twigs, sticks and logs from a variety of plant species.

A comparison of phytolith residues of test campfires made by students (who were not told what the nature of the study was) in savanna-like surroundings and of stumps that had been experimentally set alight, confirmed that the former produced a heterogeneous collection of phytoliths, while the latter left a residue that tended to be homogenous.

Rowlett and his associates noted another difference between the test fires in which logs and stumps were ignited and those in which campfires had been made: the former left residues with irregular shapes, while the latter left basin- or lens-shaped residues.

Three of the four discolored patches at Koobi Fora's FxJi-20 E site were basin-shaped, while a fourth, situated some distance from the others, was irregular. The basin-shaped patches were found to contain even more kinds of phytoliths than the test campfires had, while the remaining, irregular one contained only one kind and was presumed, accordingly, to have been made by a naturally ignited tree.

Swartkrans and Koobi Fora are not the only sites at which researchers have found possible evidence of early-Pleistocene use of fire by hominins. Hominin fire use may also have discolored the clumps of earth discovered in Chesowanja in Kenya, which were also dated to about 1.6 million years BP.

Israel has provided a link between these early African sites and later European evidence of hominin fire use. Burned seeds, wood and flint found at Gesher Benot Ya'aqov in the northern part of that country suggest that humans were, nearly 790,000 years ago, controlling fire at this site. The diverse collection of plant material burned there includes three species that had edible fruits or seeds: olive, wild barley and wild grape. The distribution of burned fragments of flint at the site was, moreover, suggestive to investigators from the Hebrew University in Jerusalem and Bar-Illan University in Ramat-Gan, that burning had occurred on specific spots, possibly suggesting the locations of hearths.

* * *

Bob Brain was careful to point out that the hominins who burned the Swartkrans bones had not necessarily learned how to *make* fire. He thought it was, instead, likely that they had harvested their fire from naturally ignited flames.

How might that "harvesting" have begun? Birds and mammals often converge on African veld fires to prey on small mammals and reptiles, as well as insects flushed out by the flames. Hominins could initially have been drawn to such fires for the same reason. Eventually, the inventive, curious natures of those beings might have prompted them to literally "play with fire" at upwind or nonthreatening edges of those burns. That could have taught them enough about the dynamics of fire to enable them to "harvest" burning material and use it to build and maintain fires at places of their own choosing.

How frequently would natural fires have occurred? In Southern Africa, the overwhelming majority of such fires burn in September and October when dry electric storms ignite grass or brush before the summer rains begin. In an article titled "Fire Behaviour in the Kruger National Park," W. Trollope and A. Potgieter report that lightning-caused fires can be expected to burn any given area in the Kruger Park at intervals ranging between five and ten years. Several subsequent studies have confirmed that a large proportion of Kruger's savanna remains unaffected by natural fire for years at a time. Daniel Koen, who has studied the issue of naturally caused highveld fires intensively to formulate a fire policy for Suikerbos-rand Nature Reserve just south of Johannesburg, tells me that, even though Suikerbosrand is one of the most lightning-prone areas in South Africa, many parts of the reserve can go for years without being burned by lightning-caused fires. The incidence of lightning strikes varies widely from year to year, and periods in which lightning-caused fires don't occur in the reserve, or affect only a small part of it, can be interrupted by a summer in which lightning strikes are widespread and frequent. Under such circumstances, successive lightning-ignited fires can start in close proximity to each other in a single season if each is extinguished by rain before it burns all the available fuel in that area.

As we've just seen though, lightning-caused fires normally take place much less frequently than this. By August of 2005 I had, for instance, been waiting two years for the veld around my favorite walking trail in Suikerbosrand reserve, Bokmakierie, to burn. (My impatience had to do with the fact that fire, followed by rain, transforms Bokmakierie into a verdant Alpine-type meadow, crowded with a well-beyond-Alpine diversity of flowering plants.) Most of the Bokmakierie area was finally burned in late 2005, whose rainy season was characterized by an exceptional number of electric storms, by one or more lightning-ignited fires, and by another started by a visitor or staff member next to a public road.

I'm impressed—especially in view of the fact that natural, harvestable fires would probably have occurred relatively infrequently—by the abundance and the ubiquity of burned bone found in Member 3. I suspect, moreover, that a great many "archaeologically invisible" fires—that is, fires that didn't contain bone—must, in addition, have been made in that Member's catchment area: bones would not always have been available to the fire users, and bone produces, besides that, acrid, unpleasant-smelling smoke when it is thrown into a fire. Taken together with the fact that fire use seems, from the investigations of Rowlett, Bellomo, Ludwig and others, to have become a widespread phenomenon in Africa around 1.6 million years BP, the high frequency of hominin-tended fires evidenced at Swartkrans prompts me to speculate that the beings who tended those fires could, by the time Member 3 was accumulating, have overcome their dependence on harvested fire and learned to kindle flames at will.

Hominins who were still dependent on harvesting fire would probably have learned, after a while, to keep their fires going for relatively long periods of time, and to transport embers over relatively long distances. Days or weeks after such harvestings, however, inattention, sleep, local fuel exhaustion, or rain would inevitably have deprived them of fire for months or even years at a time. Each of those losses would have been a bitter disappointment because the harvesters would, by that time, have experienced the comfort—and the enormous increase in personal safety—that fire could bring to their nights. Finding a way of kindling a flame would, therefore, have become a very pressing issue indeed.

Would the average *erectus* have been able to find such a way? No more, I imagine, than the average *sapiens* would have been able to invent the polymerase chain reaction in the late twentieth century. The PCR, a cornerstone of the biotech industry, makes it possible to multiply strands of DNA to large quantities for a great many scientific, medical, forensic and industrial purposes. It earned Kary Mullis, who invented it in 1981, a Nobel Prize, the prestigious Japan Prize, and a place in the National Inventors Hall of Fame. In 1991 Hoffman-La Roche bought the patent for this process from Mullis's former employer for $300 million—possibly the highest price ever paid for a piece of intellectual property.

Acheulean tools, which were coming into use during the time in question for purposes as disparate as digging in the earth, butchery, leatherwork, woodwork and scything soft plant material, attest to relatively high levels of ingenuity. So does the harvesting, transportation, and maintenance of fire. It doesn't seem unlikely to me, therefore, that one or more *erectus* counterparts of Kary Mullis could have noticed, somewhere between 1.7 and 1.5 million years ago, that sparks struck while knapping certain kinds of stone could fall, still glowing, onto dry organic material and set it smoldering, and that he (or she or they) could have exploited that state of affairs to make the discovery that Charles Darwin regarded as "the greatest ever made by man, excepting language."

* * *

It takes a great deal of solar energy to synthesize the carbohydrates—the sugars, cellulose and lignins—that go into the making of a tree. Enormous benefits would have accrued to the hominins who learned how to unlock the energy packaged into those carbs by means of setting fire to wood. Gaining that knowledge would have enabled them to literally release the light and warmth of the sun in the middle of the night—to stave off cold, say, or to repel predators. Fire could be used, too, to drive prey animals and, in some circumstances, kill them. Its heat could also neutralize bacteria and toxins, release nutrients and make hard-to-chew foods like roots, bulbs and corms palatable. Our family's fire-making ability would not, therefore, only have greatly increased our safety and comfort but also have opened up a smorgasbord of new nutritional options.

THE RELEASE OF THE HUMAN FAMILY FROM THE CONSTRAINTS THAT FORMERLY LIMITED ITS ECOLOGICAL IMPACT

For some 4 million years after our earliest ancestor split off from the chimpanzee line, it and its descendants lived in a wilderness ecology that maintained biodiversity the way that system had always done, namely by limiting the populations—and, therefore, the ecological impacts—of each of the myriad of species comprising it. Just over 2 million years ago, however, near the start of the Pleistocene Epoch, at least one member of what had by now become a family of human-like beings, evolved enough inventive intelligence to overcome some of the checks on its ecological impact. This close-to-human being's newfound ability to evade these ecological restraints led to the extermination, not long after the middle of the Pleistocene, of a number of the African animals it competed with and preyed on. Those victims were to include, by the end of the Pleistocene, the elimination and/or absorption, in both Africa and Eurasia, of at least six other members of the human family itself.

As its emancipation from the controls that formerly constrained its ecological impact escalated toward the end of the Pleistocene, our inventive being, now become *Homo sapiens*, extended its range beyond Africa into all the earth's other habitable continents. This resulted in the extermination by our species of the vast majority of the big mammals, reptiles and birds of Australia, Eurasia and the Americas.

After the Pleistocene gave way to the Holocene some 11,650 years ago, humans continued to cause a stream of extinctions as they colonized islands in the earth's seas and oceans, and developed agriculture, animal husbandry and industry. Over the past fifty years that stream has swelled into the flood of species losses presently diminishing the planet's biodiversity.

CHAPTER 14

Metaphorical New Zealand

WHAT IS IT ABOUT *HOMO SAPIENS* THAT IS MAKING IT INFLICT THIS mass extinction on the biosphere that brought it forth?

Many people have concluded, understandably enough, that a great evil of this kind could only have been perpetrated by a species that was itself evil in some sense. Raymond Dart, the brilliant, quirky anatomist-paleontologist we met earlier in this book, felt that the "loathsome cruelty" of "man" explained why "[h]e has either decimated and [*sic*] eradicated the earth's animals, or led them as domesticated pets to his slaughterhouses."

Yuval Harari, an Israeli history professor and author of the best-seller *Sapiens: A Brief History of Humankind*, also sees us as cruel beings:

> Most top predators of the planet are majestic creatures. Millions of years of domination have filled them with self-confidence. Sapiens by contrast is more like a banana republic dictator. Having so recently been one of the underdogs of the savannah, we are full of fears and anxieties over our position, which makes us doubly cruel and dangerous. Many historical calamities, from deadly wars to ecological catastrophes, have resulted from this over-hasty jump.

Harari's idea that "most top predators" are "majestic" and "self-confident" doesn't constitute serious biological thought. It's reminiscent, rather, of the medieval bestiaries in which "noble" and "merciful" lions spare women, children and released prisoners on their way home. It is, in short (as Wolfgang Pauli said about another idea that didn't merit serious discussion), "not even wrong." Important for our purposes, however, is that Harari's idea that humans are crueler and more dangerous than other animals is widely believed in contemporary society.

In its milder forms, the "humans bad, animals good" idea can sound so reasonable, so obviously *right,* that it can seem perverse to question it. Who could object, for instance, when the Discovery Channel informs us that wolves "do not exterminate other animals like humans do, because they kill only to eat"? Wolves do not, however, kill "only to eat." When abnormal snow depths on Michigan's Isle Royale and in Northern Minnesota make it relatively easy for them to kill moose and deer, they can kill up to three times as many animals as they do in a normal winter, with little or nothing eaten from some of the carcasses. Hans Kruuk's study of spotted hyenas confirmed that other predators can also, when circumstances permit, act like the proverbial fox in the henhouse:

> [I]n November of 1961 a group of hyenas killed, in one night, more than 110 Thomson's gazelle and maimed many others, eating small parts of only a few victims, (13 in a sample of 59). The carnage had taken place on a very dark night with heavy rain, and the hyenas had been able to kill at leisure.

Lions have also been observed to kill more animals than they can consume during the concentration of migratory wildebeests in East Africa's rainy season. Circumstances in which hunting becomes easy and risk-free can induce predators ranging from orcas to snow leopards to act like that proverbial fox.

Under normal conditions it's quite true, of course, that animal predators *don't* kill in such wasteful ways. This is not, however, because they are prudent, restrained or compassionate. It is so, instead, because, under the "normal" conditions we're talking about, hunting isn't easy, and involves a significant expenditure of energy on a venture that can end in failure, injury or death. But—and this is a very big but—even though predators like lions *do*, therefore, under normal conditions, usually kill "only what they need," that is still not enough to prevent them from exterminating the animals they prey on. If lions were able to kill *whenever* they needed to eat, then, even if they *always* killed "only what they needed," they would never starve. As a result, many more cubs would be born and grow to adulthood, and there would be a sudden and explosive increase in the lion population.

So why doesn't this happen?

In the early days of wildlife protection in South Africa near the start of the twentieth century, rangers in the Sabi Game Reserve were under orders to protect "game" animals by shooting predators like lions. (Leopards, cheetahs, hyenas, wild dogs and eagles were also shot.) When the modern concept of a permanent, undisturbed wilderness area emerged with the proclamation of a much-enlarged Sabi protected area as the Kruger National Park in 1926, the policy of controlling predator populations was dropped. There was plenty of opposition to this decision. Farmers lobbied against the protection of lions for the same kinds of reasons that American farmers complained about the reintroduction of wolves into Yellowstone National Park in 1995. Interestingly enough, there were also *conservationists* who opposed the extension of protection to predators. They feared it would lead to uncontrolled growth in their numbers, which would lead in turn to the disappearance of their prey species such as antelopes, giraffes and zebras.

These fears proved to be unfounded. The Kruger Park's first warden, James Stevenson-Hamilton, observed no reduction in prey populations when the policy of controlling the populations of lions and other predators was abandoned. Instead, he noticed that

> ...younger lion were often in poor, even emaciated condition, that cubs were dying, and that the number of lion of all ages, especially young animals, slain by their companions, was greatly increasing.

Putting a stop to the human control of the lion population was enough to shift that function to nature, and nature (who "understands her business better than we do," as the philosopher Montaigne pointed out more than 400 years earlier) did a thorough job, never allowing the lion population to reach a level that threatened to wipe out any of its prey species.

Today, almost a century after human control of their population was abandoned, lions still have zero population growth in the Kruger Park. More precisely, short-term increases in their population are canceled out, over time, by decreases. A lion cub born at the beginning of an El Niño drought when thousands of buffalo are starting to weaken and die is, for instance, a fortunate little creature indeed. The hunting is relatively easy in those circumstances, and the cub's chances of surviving into adulthood are comparatively high. Another which is born, on the other hand, after normal rainfall has resumed, when the reduced buffalo population that has survived the drought is getting back into peak condition, faces a bleak prospect. The stability of Kruger's lion population is not, in short, the kind of stability we associate with peace and security. It is maintained, on the contrary, by the fact that the starvation and starvation-related killings described by Stevenson-Hamilton have continued, at varying levels of intensity, up to the present day. Fritz Eloff, who investigated lion mortality in Southern Africa's Kalahari Desert, estimated that one out of every sixteen lion cubs made it to adulthood. George Schaller, working in East Africa, came up with an estimate of one in five.

Stevenson-Hamilton's observations provide us with a concrete illustration of the fact that lions don't limit their impact on the populations of their prey by killing "only what they need" but that their impact is limited, instead, by starvation in their own ranks and by conflict caused by the threat of starvation.

When I was starting to write this book in the Kruger's research camp at Skukuza, I drove off, one morning, in search of a black rhino that had been sighted near the camp. I didn't find the rhino but encountered, instead, a male lion whose skeletal appearance showed clearly that he was in the final stages of starving to death. As I watched him walking slowly along the road next to my truck, indifferent both to me and to the herds of equally indifferent impala antelopes we were passing, I realized that if I were a ranger stationed in one of the remote areas of this vast protected wilderness, I might well have shot one of those impalas for him, but that would, of course, only have postponed the inevitable. Maybe, I thought, I would have put the lion out of his misery instead—I certainly wouldn't have let any of the cats I've owned suffer like that. Here I was, anyway, fantasizing, despite all the fascination and reverence that wilderness ecologies produce in me, about curtailing or mitigating the workings of this one.

Lions don't, of course, lead lives of unremitting desperation, and it's difficult to remember the harsh realities that control their population when you're looking at an obviously well-fed pride relaxing next to one of Kruger's roads. As Darwin wrote in the *Origin*:

> We behold the face of nature bright with gladness, we often see superabundance of food; we do not see, or we forget, that the birds which are idly singing round us mostly live on insects or seeds, and are thus constantly destroying life; or we forget how largely these songsters, or their eggs, or their nestlings, are destroyed by birds and beasts of prey; we do not always bear in mind, that though food may be now superabundant, it is not so at all seasons of each recurring year.

Lions, like literally all other life-forms, produce more offspring than their environment can support. It's hard to see how nature could have arranged things differently. How could living organisms have evolved, after all, to produce only the right number of offspring to replace their parents rather than more, or any less? How can there, in fact, even *be* a predictably "right" number of offspring of any species in the interdependent, feast-and-famine world of all the life-forms constituting a wilderness ecology? How could that "right number" be knowable in advance in a biosphere that wasn't completely predictable and accident-free?

If we accept that the consistent production of the "right" number of offspring is an impossible task, then only two other possibilities remain:

1. a tendency to produce less offspring than are required for replacement, or

2. a tendency to produce more.

Possibility 1—less offspring—is clearly out of the question: organisms whose reproductive capacity is limited in this way would obviously become extinct fairly quickly.

We're left, therefore, with the seemingly inescapable conclusion that the only chance of a species staying on as a member of our biosphere for anything more than a trivial amount of time lies in its consistently producing *more* offspring than the amount "usually" or "normally" required to replace the parent generation.

That conclusion is supported by the observable facts: every known species, from codfish (each of whose females may produce millions of eggs in a single year) to gorillas (whose females generally give birth to a single infant every four years), *does* produce more offspring than are required to replace the parent generation. There is only one partial and recent exception to that state of affairs: some members of the human species—paradoxically the very species whose population has quadrupled in the last hundred years—have begun to limit their reproductive activities voluntarily.

* * *

Even people who profess to accept Darwin's theory of natural selection can have difficulty with the hard core of that theory—that is, that the population of every organism on earth—from viruses to intelligent, slow-breeding animals like dolphins—is limited by collisions with the wall of limited resources. Darwin himself found it hard to accept and apply this idea consistently:

> Nothing is easier than to admit in words than the truth of the universal struggle for life, or more difficult—at least I have found it so—than constantly to bear this conclusion in mind. Yet unless it be thoroughly engrained in the mind, I am convinced that the whole economy of nature, with every fact on distribution, rarity, abundance, extinction and variation, will be dimly seen or quite misunderstood.

One of the reasons that the real facts about evolution and natural selection remain, in the words of Richard Dawkins, "largely unabsorbed into popular consciousness" is that people find it hard to accept that unpleasant things like starvation, frustration and conflict have always limited wild animal populations. It is even harder to accept that they have, until very recently, also limited the growth of the human population. Look, for instance, at how completely a popular writer on human evolution like Robert Ardrey misunderstood this function of the ecology in his African Genesis:

> ...dominance over-developed can do damage of an absolute order upsetting natural balances otherwise so carefully protected. The appalling death rate of juvenile lions, for example, as recorded in the Kruger reserve, can scarcely be regarded as in the long-term interests of natural selection or in the short-run interests of the pride.

Darwin came to understand, of course, that both natural selection and biodiversity rested foursquare on those "appalling" death rates, but he found that fact difficult to reconcile with the idea of a benevolent God.

Darwin was a compassionate man. The only serious argument he had with Captain Fitzroy of the *Beagle* during their five-year voyage around the world involved slavery. Early on in the voyage, on one of their stops along the coast of Brazil, a landowner had, in the presence of Fitzroy, asked his slaves how they felt about being slaves. The slaves had, according to Fitzroy, assured their master that they were satisfied with their lot. Fitzroy told Darwin about this incident, suggesting that it showed that slavery wasn't as bad as abolitionist propaganda was trying to make out. Darwin, whose family had been on the forefront of the anti-slavery movement for decades, replied ("perhaps," he admitted, "with a sneer") that it was unrealistic to expect anyone to express his real feelings about his situation in the presence of a master who had the power of life and death over him. This answer provoked such a furious response from Fitzroy that Darwin feared that he might be ordered to leave the *Beagle*. After some delay, however, Fitzroy apologized for his outburst, and Darwin's history-making voyage was allowed to continue.

* * *

Darwin wasn't only compassionate—he was also religious. When he started his voyage, both he and his family were expecting him to become a clergyman on his return to England.

In addition to this conventionally religious aspect of his life, Darwin had a spiritual experience of a more personal kind when he arrived in South America as a young man of twenty-three. After walking by himself in a South American rainforest for the first time on February 29, 1832, he wrote in his diary that "[I]t is not possible, to give an adequate idea of the higher feelings of wonder, admiration and devotion which fill and elevate the mind..." So powerful was the experience that Darwin felt that its intensity might never be equaled:

> To a person fond of natural history, such a day as this, brings with it a deeper pleasure than he ever may hope again to experience.

The passage of time confirmed the uniqueness of Darwin's experience. Reworking his diary seven years later for publication under the title *Journal of Researches into the Geology and Natural History of the Various Countries Visited by HMS* Beagle, he changed the phrase "a deeper pleasure than he ever *may* hope again to experience" to "a deeper pleasure than he ever *can* hope again to experience." Commenting more than forty years later on the profound feelings he had experienced "standing in the midst of the grandeur of a Brazilian forest," he wrote in his *Autobiography* that "I well remember my conviction that there is more in man than the mere breath in his body."

That conviction was not destined to last. It's not particularly surprising that Darwin became an atheist—his father, grandfather and older brother Erasmus had, after all, preceded him along that road. He described his loss of faith with the same ingenuous honesty that characterized his scientific observations:

> ...I was very unwilling to give up my belief;—I feel sure of this because I can well remember often and often inventing daydreams of old letters between distinguished Romans and manuscripts being discovered at Pompeji or elsewhere which confirmed in the most striking manner all that was written in the Gospels. But I found it more and more difficult, with free scope given to my imagination, to invent evidence which would suffice to convince me. Thus disbelief crept over me at a very slow rate, but was at last complete.

It wouldn't be true to say that Darwin's discovery of the phenomenon of natural selection *caused* this growing disbelief (any more than his lasting grief about the death of his beloved nine-year-old daughter Annie caused it). It is true, however, that he found the idea of a struggle for existence intellectually easier to reconcile with atheism than with the existence of a benevolent God. "That there is much suffering in the world," Darwin wrote in his *Autobiography,* "no one disputes."

> Some have attempted to explain this in reference to man by imagining that it serves for his moral improvement. But the number of men in the world is as nothing compared with that of other sentient beings, and these often suffer greatly without moral improvement. A being so powerful and so full of knowledge as a God who could create the universe, is to our finite minds omnipotent and omniscient, and it revolts understanding to suppose that his benevolence is not unbounded, for what advantage can there be in the sufferings of millions of the lower animals throughout almost endless time? This very old argument from the existence of suffering against the existence of an intelligent first cause seems to me a strong one; whereas, as just remarked, the presence of much suffering agrees well with the view that all organic beings have been developed through variation and natural selection.

Hard as it may be to accept the idea of an ecology that tolerates suffering, we have to go a step further to get to a full understanding of Darwin's insight: the ecology doesn't simply *tolerate* frustration, conflict and starvation. It actually *requires* those things—depends on them—to maintain its integrity. A large proportion of all the organisms that inhabit and constitute a wild African savanna *must* die of something other than old age and/or have their sexual urges frustrated if the biodiversity of that savanna is to be preserved.

Schopenhauer's assertion that "...we require at all times a certain quantity of care or sorrow or want, as a ship requires ballast, in order to keep on a straight course…," may be debatable in the philosophical context in which he wrote it, but it's a disconcertingly accurate description of the workings of a wilderness ecology. "If you're not shocked by the theory of quantum mechanics," Niels Bohr, the Danish physicist, was supposed to have said, "then you can't have understood it." One could paraphrase Bohr's aphorism by saying that if you're not dismayed by the way natural selection works, then you haven't yet understood what *that* doctrine is about.

Scientifically trained people are as likely as anyone to react to the "dismaying" part of Darwin's discovery by de-emphasizing or denying it. The Canadian environmentalist David Suzuki provides us with an example of that kind of denial:

> Scientists are first and foremost human beings, as subject to cultural bias as anyone. So it should not be surprising that Charles Darwin, educated in a period of slavery, colonialism and mercantile expansion, should have described his great insight—evolution by natural selection—in terms of struggle and selection of the fittest.

Suzuki seems to be suggesting that Darwin might have described his "great insight" in terms of cooperation rather than a struggle for existence if he'd been born in a less rapacious time. Cooperation is, after all, a common phenomenon in the natural world. The many forms of cooperation found in nature are, however, as solidly based on the realpolitik of self-interest as the "murders" and "deceptions" the non-cooperating members of the biosphere commit to survive and reproduce. Certain plants, for instance, play an important role in helping wasps to find their caterpillar victims by secreting volatile substances that alert the wasps to their presence, but they only do so when the caterpillar is eating their leaves. "Myrmicophylic" or "ant-loving" caterpillar species may, in their turn, send out alarm signals (in both sound and pheromonal form) when they are threatened by wasps. Those alarms summon ants to their defense, but the ants only drive the wasps away because of the interest they have in the nectar that the caterpillar secretes for their benefit.

Cooperation does not, therefore, constitute an alternative or an exception to the primacy of self-interest in the natural world. In his *Climbing Mount Improbable* Richard Dawkins tells us, for instance, that the "pacts" that have established a mutually dependent reproductive strategy between almost every one of the world's 900-plus fig species and a particular wasp species peculiar to almost every one of them are "redolent of hard bargaining, of trust and betrayal, of temptation to defect policed by unconscious retaliation."

Cooperation is without question a fundamental part of the biosphere, extending right down to organelles like the mitochondria in our cells, which have their origin in the symbiotic association of two unrelated life-forms. It seems, however, quite as likely to promote competitive struggles as it is to supplant them. Wolves would not, for instance, have the opportunity of preying on animals as big as adult moose if they hadn't developed the cooperative behavior required to hunt in packs. Warfare itself (whether between ant or primate groups) could scarcely take place in the absence of highly developed cooperative behaviors.

Despite the fact that cooperation plays an important role in the biosphere—and indeed partly *because of* that fact—the natural world continues to be dominated by the struggle of living beings to find their place in a world that cannot allow them all to survive and reproduce. As John Maynard Smith and Eörs Szathmáry explain in their *Origin of Life,* the countless organisms in our biosphere that cooperate with each other wouldn't exist in the first place if it weren't for the struggle for survival and reproduction:

> Lynne Margulis, who marshalled the evidence that persuaded biologists that mitochondria and chloroplasts were once symbionts, has sometimes argued that symbiosis is the main source of evolutionary novelty, and that natural selection has been of

minor importance. This will not do. Symbiosis is important because both partners contribute something. In nitrogen-fixing
symbiosis, for example, Rhizobium [a bacterium] contributes
the ability to fix nitrogen, and the plant contributes photo-
synthesis and the whole anatomy of root and shoot to succeed
on land... These are complex adaptations that could only have
evolved through natural selection. The motor bike is a symbiosis between the bicycle and the internal combustion engine.
It works fine, if you like that kind of thing, but someone had to
invent the bicycle and the internal combustion engine. Symbiosis is not an alternative to natural selection: rather, we require a
Darwinian explanation of symbiosis.

When he suggested, therefore, that Darwin's "great insight—evolution by natural selection" didn't have to be described "in terms of
struggle and selection of the fittest," David Suzuki might just as well
have been telling us that Copernicus's "great insight" didn't have to
be described in heliocentric terms.

* * *

"Heaven and Earth are heartless," the *Tao Te Ching* tells us, "treating
creatures like straw dogs." Which organism wouldn't, if it had the
power to do so, commandeer a bigger share of the resources that a
wilderness ecology doles out with such tender solicitude for biodiversity and such callous disregard for the preservation of individual
lives? With more understanding of her situation, a human living
200,000 years ago might have wished, therefore, to escape not just
from *this* leopard, *this* murderous member of her own species, or the
effects of *this* dry season, but from the entire wilderness ecology—
we could call it "the Metaphorical Serengeti"—that made her life,
and the lives of her family, so unpredictable and insecure.

Other species have enjoyed temporary and local escapes from
the limitations imposed by this Metaphorical Serengeti. When a
handful of rats were, for instance, brought as stowaways (or stored

food) to New Zealand by its Polynesian discoverers in the four-teenth century, they bred explosively to form what Tim Flannery suggests might have been "...plagues of such vastness that they have never since been rivalled." There's nothing mysterious about these plagues: it's quite predictable that a rat population, freed from the army of avian, reptilian and mammalian predators that kept it in check on the Asian mainland and presented with a seemingly limit-less source of naive prey, would experience that kind of explosive increase. Also predictable was the crash that must have occurred when these prey species (flightless or weakly-flighted, ground-dwelling birds, short-tailed bats that did much of their hunting on the forest floor, six species of endemic frogs, a rich radiation of geckos and skinks, large endemic snails and giant cricket-like insects) were driven into scarcity, or—as actually happened in most cases—outright extinction.

The fact that humans and their domesticated livestock now account for about 96 percent of the biomass of all mammals, while wild mammals have been reduced to only 4 percent, invites a conclu-sion that may seem unlikely but is, at the same time, completely obvious: like New Zealand's first rat immigrants, *Homo sapiens* has temporarily overcome the diversity-preserving limitations we've been discussing in this chapter. In the next chapter we'll discuss the "radically new faculty" that allowed us to make that escape, and enter the unprecedented Metaphorical New Zealand we occupy for the time being.

CHAPTER 15

A Radically Different Kind of Faculty

Rɪᴅɪɴɢ ᴀ ᴘᴏɴʏ ɪɴ ᴛʜᴇ ᴍᴏᴜɴᴛᴀɪɴᴏᴜs ᴄᴏᴜɴᴛʀʏ ᴏꜰ Lᴇsᴏᴛʜᴏ, I became nervous when the narrow, rock-strewn trail we were negotiating skirted the edges of steep precipices. I found myself paying close attention, at such places, to where my horse was placing its hooves. On several occasions, I saw it test footholds, before shifting its full weight onto them. These were reassuring acts. My animal had the "horse sense," I realized, to choose "wise" or "judicious" ways to negotiate that potentially hazardous trail, without possessing anything like a human level of intelligence. Nature had, it seemed, equipped it with neural machinery dedicated to the production of sophisticated all-terrain mobility. That machinery may have been cheap and simple compared to the more general and abstract kind of intelligence our species possesses, but I began to realize to my relief that it seemed to be reliable.

Natural selection is every bit as capable of assembling complex behaviors and abilities as it is of producing complex body shapes. Consider, for example, the inborn "promptings" and "guidance" that will allow a young Arctic tern to make its first migratory flight from northern Canada down to Antarctica.

The bird's parents will leave before it does, so it has to rely entirely on a genetic "navigation package" to initiate and complete that 10,000-mile journey. Ingenious experiments relating to the directions in which caged birds orient themselves in planetariums when they're in the grip of *Zugunruhe*—migration unrest—have shown that some migratory bird species have "star maps" written into their DNA, and it's possible that Arctic terns are also equipped with "instinctive" sky charts of that kind. It's difficult to imagine how chance variations—each separately advantageous to its possessor—could have accumulated to put together that kind of map, as well as the "clocks," "magnetic compasses" and other specialized structures that may constitute the tern's "avionics," but natural selection is, as Richard Dawkins's 1997 *Climbing Mount Improbable* demonstrates so convincingly, equal to tasks of that kind.

While the mills of natural selection can, therefore, grind exceedingly fine, they also grind very *slowly*. Bats took millions of years to develop the "hardware" and "software" required to send out the sound-pulses whose echoes locate the insects they feed on, while the insect species that execute spiraling crash-dives when they hear those sound-pulses, took similarly big chunks of time to evolve that defense. If insects were able to understand what was happening when they heard bats' "sonar" pulses, then they could have adopted countermeasures such as the crash-dive immediately. Because they don't have that kind of understanding, the impulse to make power-dives in response to sonar pulses had, of course, to be constructed entirely by the accumulation, over hundreds of thousands of generations, of chance variations in the reactions of individuals of their species to those pulses.

Animals of all kinds are equipped to react to particular situations in such uncomprehending but appropriate ways. Humans themselves don't have to understand that spiders can give them dangerous bites or that contact with excrement can pass contagious diseases to them. Specialized neural structures tell us instead—in a peremptory way independent of both learning and logic—that spiders look frightening and that excrement is disgusting. Much of what we think and do is powerfully affected by instincts. We don't have to be taught to fear heights, enjoy sweet tastes, find someone beautiful or seek the respect of our fellow humans. A multitude of instincts shape our behavior with such effortless power that we tend to be blind to their existence.

> "Why," William James asked, "does a particular maiden turn our wits so upside-down? The common man can only say, "*Of course* we… love the maiden, that beautiful soul clad in that perfect form, so palpably and flagrantly made from all eternity to be loved!"

It takes what James calls "a mind debauched by learning" to understand that our fascination with the maiden (or her male counterpart) is a product of purpose-built brain regions whose operations are triggered by criteria installed in our minds by natural selection for reproductive purposes—criteria as meaningless to other species as their "attraction triggers" are to us. "What voluptuous thrill," James goes on to ask,

> may not shake a fly, when she at last discovers the one particular leaf, or carrion, or bit of dung, that out of all the world can stimulate her ovipositor to its discharge? Does not the discharge then seem to her the only fitting thing? And need she care or know anything about the future maggot and its food?

Strokes and other cerebral disruptions can bring home to us in a dramatic way that ostensibly "simple" or "natural" activities like walking and eating are, like sexual attraction, the elaborate productions of specialized neurological systems. Even the moral dimension of our species' existence, often thought of as a kind of opposite pole to the instinct-dominated, "animal" part of our make-up, originated in, and is still influenced by, the kind of "special-purpose" neural structures that characterize instinctive behavior.

"I fully subscribe," Darwin wrote in *The Descent of Man, and Selection in Relation to Sex*, published in 1871, "to the judgement of those writers who maintain that of all the differences between man and the lower animals, the moral sense or conscience is by far the most important." He goes on, however, to point out that despite the fact that morality appears to distinguish us from the "lower animals," its origins also lie in natural selection:

> The first foundation or origin of the moral sense lies in the social instincts... Animals endowed with the social instincts take pleasure in one another's company, warn one another of danger, defend and aid each other in many ways. These instincts do not extend to all the individuals of the species, but only to those of the same community. As they are highly beneficial to the species, they have in all probability been acquired through natural selection.

From this he concluded that

> any animal whatever, endowed with well-marked social instincts, the parental and filial affections being here included, would inevitably acquire a moral sense or conscience, as soon as its intellectual powers had become as well, or nearly as well developed, as in man.

This argument takes the theory of natural selection into what some might see as the inner sanctum of humanness. Even today, a century and a half after it was made, it can still present an unsettling challenge to religious conceptions of morality. "Your father's opinion that all morality has grown up by evolution," Emma Darwin wrote to her son Francis after her husband's death, "is painful to me."

*　*　*

No human urge appears, then, to be too complex—or indeed too "exalted"—to be the product of evolution by natural selection. Evolution can, as we've seen, construct instinctive behaviors whose sophistication strains our credulity. "There is simply no denying," Daniel Dennett declares in his *Darwin's Dangerous Idea,* "the breathtaking brilliance of the designs to be found in Nature":

> Time and time again, biologists baffled by some apparently futile or maladroit bit of bad design in nature, have come to see that they have underestimated the ingenuity, the sheer brilliance, the depth of insight to be discovered in one of Mother Nature's creations. Francis Crick has mischievously baptized this trend in the name of his colleague Leslie Orgel, speaking of what he calls "Orgel's Second Rule: Evolution is cleverer than you are".

"Orgel's Second Rule" is—at least at this stage of our intellectual development—unassailable. The level of "ingenuity" that has gone into the assembly of the instinctive behaviors we've been talking about still exceeds that of our species by a wide margin. And yet the world of instinctive behavior has limitations. Such behaviors can only evolve in response to situations that arise over and over in the life of a particular species for a great many generations. Innovation by means of natural selection simply isn't flexible enough to respond in a creative way to the enormous world of opportunities that don't present themselves in this repetitive, "stereotyped" way.

If an organism could, therefore, retain the benefits conferred on it by instinctual behaviors, but develop, in addition, a method of devising useful responses to opportunities *lying beyond the reach of the instinct-building mechanism,* it would gain access to a cornucopia of new strategies to maximize its survival and reproduction. The human family has developed such a method. It did so by evolving "general purpose" computational machinery that can *invent* useful responses to this previously unexploitable class of opportunities.

* * *

Because inventions are "thought up" by individual members of our species, they're referred to as "ontogenetic" innovations from the Greek word "ὄντος" ("ontos") meaning "individual." The stone cutting-tools that made their first appearance in hominin history on the other side of 3 million years ago were products of this "ontogenetic" kind of innovation. The meat-cutting or "carnassial" teeth of lions, which shear against each other like sharp scissors, are, on the other hand, "phylogenetic" innovations because they were produced, over tens of millions of years, by the lion's φῦλον (phylum, or "kind") by way of evolution by natural selection.

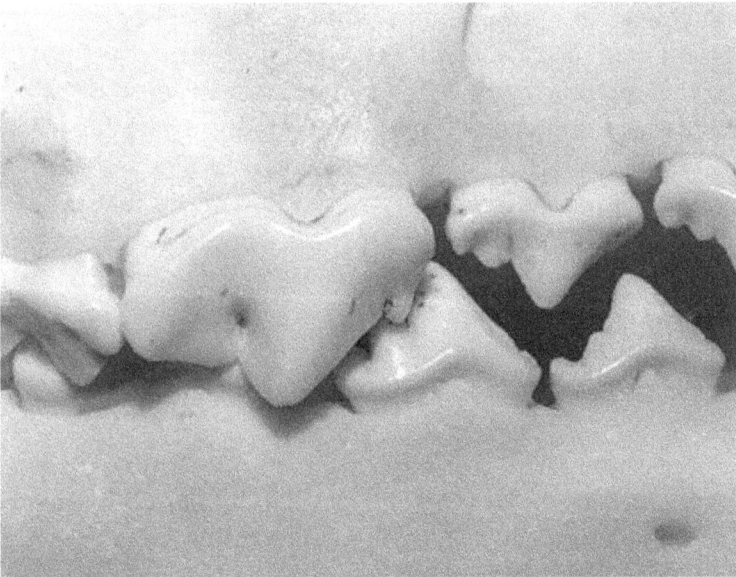

Stone cutting-tools were produced, of course, by a *series* of discoveries or inventions, rather than by a single "bright idea." Among the first of their many inventors may have been the hominin or hominins who realized that sharp flakes, accidentally chipped off stones being used as hammers, could be useful. These pioneers would have been followed by other innovators who learned to produce sharp flakes of this kind deliberately, and still others who found out how to sharpen and shape them by "pressure-flaking" their edges with hard pieces of horn, bone or wood.

Because the beautifully shaped stone cutters, projectile heads, cleavers and awls of the late Paleolithic were the products of a multi-generational enterprise stretching back into the Pliocene, we can legitimately talk about the "evolution" of stone tools. In this context, however, we're not talking about the kind of evolution that Darwin and Wallace discovered—that is, the natural selection of physical or mental characteristics thrown up by genetic happenstance. Each of the innovations, big or small, contributing to this "evolution of stone tools," would, on the contrary, have been the result of a useful idea that dawned on an individual member of the human family. Some of those insights would have resulted from the concentrated kind of care and attention we call "work," while others may have flashed into the inventor's consciousness in the course of frivolous activity or seeming idleness. But each of them—even the ex post facto recognition of the usefulness of a happy accident—would have been the active creation of a particular hominin brain.

Around the time of Darwin's birth in 1809, Jean-Baptist Lamarck had already figured out that organisms evolve into other organisms. He thought, however, that a "natural tendency toward progression and increasing complexity" was involved in this process, and that this tendency was driven, at least in part, by the transmission of acquired characteristics: an individual animal developing an ability to run particularly fast by regular and determined practice, could, he thought, pass that ability to its offspring. These assumptions turned out to be wrong. Organic evolution—what we've been talking about as "phylogenetic innovation"—proceeds, as Darwin and Wallace would show, by the natural selection of randomly appearing variations that can, but do not necessarily, bring about "progression" or "increased complexity."

The other kind of evolution we've been talking about—technological evolution—*is*, however, driven by the transmission of acquired characteristics. Bright ideas—the devices or methodologies we've been calling "ontogenetic innovations"—are, of course, acquired by the inventor during his or her lifetime. In a sense, therefore, ontogenetic innovation is a Lamarckian process, except that the transmission of the innovation is not restricted, as it is in the Lamarckian scheme, to the innovator's descendants. Neither Wilbur nor Orville Wright had any children, but they didn't of course need any to pass their invention of powered, heavier-than-air flight on to their fellow humans.

* * *

While ontogenetic and phylogenetic innovation are fundamentally different processes, neither could have operated in complete isolation from the other. A small increase in manual dexterity could, for instance, have stimulated simple ontogenetic innovations like the modification of twigs or sticks for use as tools, which could, in turn, have stimulated the natural selection of further physical and mental capabilities and so on, in an ongoing cycle of reciprocal augmentation. The control our ancestors achieved over fire more than 1.5 million years ago was a giant leap for ontogenetic innovation, but, as Richard Wrangham points out, its phylogenetic "collaborator"—that is, natural selection—may have responded with significant changes of its own. Because our tough, often fibrous savanna foods were now being softened and pre-digested by cooking, he argues, natural selection was able to reduce the size of our jaws, chewing muscles, teeth and colons, permitting the calories saved in these ways to be used for other purposes, such as feeding our energy-hungry brains and enabling the development of the bigger, stronger bodies.

Ontogenetic and phylogenetic innovation may have remained more or less equal partners in the business of creating new hominin capabilities during the Pliocene and early Pleistocene, but, toward the end of the latter epoch, a host of new productions like painted and carved images, flutes, fat-burning lamps, rope, needles and threads, together with new weapons like atlatls and bows, make it clear that ontogenetic innovation was increasing the power of our species at a speed far exceeding the workings of natural selection, giving *Homo sapiens* an overwhelming advantage over other life-forms well before the invention of agriculture.

* * *

When *The Origin of Species* appeared in 1859, there were only about 1.25 billion people on the planet, and the human impact on the biosphere would not have become anywhere near as obvious as it has in our time. Under those conditions, Darwin could still write that the opposing forces in his "war of nature" are "so nicely balanced that the face of nature remains uniform for long periods of time." Darwin recognized that our species had become "the most dominant animal that has ever appeared on the earth" and that our power of invention was an important contributor to that state of affairs:

"He [man] has invented and is able to use various weapons, tools, traps, etc., with which he defends himself, kills or catches prey, and otherwise obtains food... He has spread more widely than any other highly organised form; and all others have yielded before him."

It could not, however, have been as obvious to Darwin as it has become to us, that the human power to invent useful new devices and behaviors is disturbing the "nice balance" keeping his "economy of nature" (termed "der Ökologie" by his admirer Ernst Häckel in 1866) within limits compatible with biodiversity. As we saw in Chapter 2, however, the psychologist Leda Cosmides and the anthropologist John Tooby realized that human power of invention is able to produce "...innovations that are too rapid with respect to evolutionary time for their antagonists to evolve defenses by natural selection..." and that the possessors of this ability would, as a consequence of this advantage, bring about "the extinction of many prey species in whatever environments they have penetrated."

* * *

I don't want to suggest that the world of ontogenetic solutions is entirely closed to life-forms who are not members of the human family. Some species have been able to respond to novel opportunities in ways that can fairly be called "inventive." A female raven caged with others of her species in temporary captivity by Bernd Heinrich in Maine, was able, for instance, to get hold of a piece of meat suspended from a perch by a two-foot length of string. The bird did so by flying to the perch, reaching down to pull some of the string up with her beak, then standing on the pulled-up length with her foot, and repeating the operation. She did not, Heinrich thinks, hit on this solution by trial and error. For several hours after Heinrich tied the meat-and-string device to the perch, none of the birds came near it. Then the "inventor" abruptly flew to the perch and used the "beak and foot" method to haul the meat up without further ado. Heinrich concludes that the solution occurred to her as the result of a mental examination of the problem.

Corvids, the group to which ravens, crows, magpies and blue jays belong, are a very smart family. New Caledonian crows modify and use twigs as tools, and in Richmond, British Columbia, I witnessed crows dropping hazelnuts that they had brought from a nearby tree onto a busy street during rush hour so that the cars' tires would break their shells. It was a startling sight to see those birds swooping down into the traffic to retrieve the kernels of those nuts—and to steal them from one another.

A Japanese macaque monkey became famous for learning to separate rice strewn on a beach from sand by floating the former away from the latter in water, a method that some members of her band were able to imitate. We've seen, too, that chimpanzees use modified twigs, sticks and unmodified stones as tools, and that South American capuchin monkeys use stones to break the shells of nuts. The world of invention and improvisation is not, however, restricted to primates and corvids or, indeed, to vertebrates. The octopus *Amphioctopus marginatus* can carry around with it the shells of fellow members of the phylum Mollusca, or a pair of coconut half shells, cut by humans, which it will fit together around it, to conceal itself from predators and/or lie in wait for prey such as crabs.

To say, however, that "both humans and some non-human organisms can invent beneficial new ways of doing things" without qualification would be as misleading as saying that "both humans and earthworms have light-perceiving organs." Earthworms don't have eyes—they can only detect changes in the intensity of light falling on the front ends of their bodies with the aid of light-sensitive cells situated in that region. Humans, on the other hand, can use light to create exquisitely precise mental images of their surroundings.

Reproduced with the kind permission of Rokus Wessel Groeneveld (www.diverosa.com).

The difference between the light-manipulating ability of humans and those of earthworms is, as I see it, a roughly accurate metaphor for the difference between the inventive power of humans and those of non-human organisms. The former is so much greater than the latter that it is, for all practical purposes, a radically different kind of faculty.

* * *

The ability to think up solutions to novel problems seems to require the ability to construct mental models of causal relationships in the real world. "What if," a member of our species might ask herself, "I take the noose I've tied into this cord and hang it across an overgrown path that small antelopes like duikers use, and then tie the other end of the cord to a branch?"

To answer that question, her brain will produce a kind of "movie" of the consequences likely to flow from this action. The plotline of that "movie" will conform to her idea of how the causal structure of the real world works. If her understanding of that causal structure is accurate enough, it will allow her to predict correctly that duikers aren't likely to spot a noose positioned across a narrow path through tall grass and that they could, therefore, put their heads into such nooses and then pull them tight in their efforts to escape.

If too many duikers manage to pull their heads out of those nooses (in either her mental movie or in reality), then sequels will appear, produced by the original snare-designer, or by others, in which young trees whose crowns are bent down and secured to the ground with hair-trigger connections, might spring upright to hoist the hapless little antelopes off their feet when those connections are disengaged by a tug on the noose.

Hominins have become so proficient at using such "movies" or "models" to evaluate possibly advantageous new behaviors that they've become the sole occupants of Tooby and DeVore's "cognitive niche." One might suppose that our duiker trapper was able to "think up" her snares because she had emancipated herself from the "primitive" instinctual functions of her brain, but that may not be the case. The "general purpose" computational machinery evolved by our species hasn't turned its back on our special-purpose neural structures—our instincts—as if they were poor relatives. It employs them, instead, and seems, in fact, to be dependent on them. Our duiker trapper's general purpose computational machinery might have been overwhelmed, for instance, by the volume and complexity of the calculations required to model the physics of the snares she was considering, if it didn't have a "cheat sheet" of "internal" or "intuitive" physics to refer to.

General-purpose or inventive intelligence is also thought to employ special-purpose instinctual modules to help avoid what cognitive theorists and specialists in the field of artificial intelligence refer to as "combinatorial explosions": the fact that even a small increase in the elements of a problem leads to exponential—that is, explosive—growth in the number of ways those elements can combine with each other. The nine-step game of tic-tac-toe can, for instance, unfold in 362,880 different ways. Computers can review all these in real time in order to avoid the ones that lead to loss, but humans can't ordinarily manage a real-time review of that size.

The world of all possible chess games is too huge to be subjected to a review of this kind by either humans or computers. (Computers beat us at chess only because they can see further into that dense bush of variations—that is, construct bigger "branches" of consequences for particular moves than we can.)

The process of human invention cannot and does not involve sorting through enormous numbers of combinations by brute force. Trying to design a duiker snare by considering all the possible ways that the raw materials in your environment can be processed and combined would be like trying to write a book on French cooking by generating all the permutations that all the letters of the English alphabet, upper and lower case, together with the common punctuation marks plus a blank space, can assume in a 200-page sequence. Although those permutations include what Daniel Dennett refers to in his *Darwin's Dangerous Idea* as a "Vast" (which he defines as "very-much-more-than-usual") number of books on French cooking, they are so Vastly outnumbered by other possible books, which are, in their turn, so Vastly outnumbered by gibberish books, that an attempt to generate even one of them by a process of random iteration, would be Vastly unlikely to succeed on a schedule relevant to you, your publisher, or indeed, the lifetime of the Universe.

Our duiker snare designer could not, in short, have succeeded in her task if her thinking wasn't shepherded toward potentially productive areas with the help of well-stocked caches of instinctual, personal and cultural information, as well as input from her biological drives and emotions. Cognitive theorists talk about this "shepherding" process with words related to steering (cybernetic) or finding (heuristic), but it remains an intriguingly mysterious process.

"Nothing is more admirable," the Scottish philosopher David Hume wrote almost three centuries ago in his *Treatise of Human Nature,*

> than the readiness with which the imagination suggests its ideas and presents them at the very instant in which they become necessary or useful. The fancy runs from one end of the universe to the other in collecting those ideas which belong to any subject. One would think the whole intellectual world of ideas was at once subjected to our view and that we did nothing but pick out such as were most proper for our purpose. There may not, however, be any present, besides those very ideas that are thus collected by a kind of magical faculty of the soul...

That "magical faculty" is still, to use Hume's own words, "inexplicable by the utmost efforts of human understanding." Its results are, however, familiar enough: while human thought processes are fallible, they can make breathtaking leaps of discovery and invention. Some of those leaps are made in at least partly conscious ways, while others may make their appearance without the intervention of conscious logic, as "revelations," intuitions, dreams and gut feelings.

* * *

For all the dangers and dilemmas it has created for ourselves and our biosphere, the "general purpose" or "abstract" intelligence we've been talking about is obviously (to its possessors anyway) an immensely beneficial tool. It may seem surprising, therefore, that in over 4 billion years of biological evolution only one family—that of the hominins—has managed to develop it.

Presumably, abstract intelligence took more time to evolve than the modules that run specific or "instinctual" mental abilities (such as my Lesotho pony's ability to negotiate precipitous mountain trails) because the complexity of the neural systems that run the former has to be of a considerably higher order than those that enable the latter. It's conceivable, too, that an abstract intelligence of our kind could not have come into existence before the "operating system" of instinctual behaviors and information on which it appears to depend reached a critical level of richness and flexibility. Speculating along similar lines more than a century ago, William James suggested that the power of the human intellect stems from the fact that our species has *more*, rather than less, instincts than other animals.

It may be a truism to say this, but our ability to engage in general-purpose computation must have begun to appear, like any other biological system, as soon as it became both beneficial and feasible for natural selection to assemble it. Having made that appearance, it has, however, given our species the power to transform the biosphere so profoundly that no other organism on this planet may get the opportunity of evolving it again.

In order to survive on the African veld 3 million years ago, our ancestors had to compete with a formidable suite of big predators and herbivores. The outcome of that struggle was by no means assured. The only weapon our early relatives possessed to counter the superior speed and strength, and the all-too-tangible teeth, claws, hooves and horns of their adversaries, was an abstract and still-rudimentary ability to think up useful new devices and behaviors.

Today we're involved in another struggle with an uncertain outcome: to prevent this power of invention—now grown to overwhelming and all-conquering proportions—from making the earth unlivable for other organisms, and, indeed, for ourselves. The only weapon available to us in this latter-day struggle is, ironically, the same one that unleashed our destructiveness in the first place: the analytical and creative power of the human brain.

Addendum

Thoughts on Saving Species and Ecosystems Threatened by Our Own Species, and the Emotional Impact of Their Destruction

The Notre Dame Cathedral is burning as I write this. This twelfth-century building with its magnificent flying buttresses is an engineering and an aesthetic marvel. Anti-clerical fanatics damaged it extensively at the time of the French Revolution, but people were able to restore it. And we're going to restore it again after this disaster. It won't be exactly the same, but our grandchildren will have the same experiences in it as those of us lucky enough to have seen it.

What if we degrade an ecosystem? Can we restore that?

South Africa's experience with the proclamation of the Sabi Game Reserve in the late nineteenth century, enlarged and renamed as the Kruger National Park in 1926, suggests that the answer is "Yes, we can, as long as the damage is relatively limited."

When I first saw South Africa's Kruger National Park as a small boy in 1954, it was an ecologically impoverished area. Of the four biggest animals that had lived there up until the late nineteenth century (elephant, hippo, white rhino and black rhino), only hippos survived. There were still elephants in the Park's remote northern end, but black rhinoceroses and their giant relatives, white rhinoceroses, had disappeared from the entire area.

I pleaded with my parents, during that 1954 visit, to drive up to Kruger's northern end so we could have a chance of seeing the elephants. Like the vast majority of visitors to the Park during the 1950s they preferred, however, to stay in the south. Looking back I suppose it could have been a matter of time and convenience for them. We would have had to drive more than a hundred miles to reach the Park's northern regions and that would have taken more time than they'd planned to spend away from Johannesburg. There was more to it than that though. A trip to the northern part of the Kruger Park was regarded, at that time, as a kind of expedition rather than a family vacation.

The feeling I got as a child was that Kruger's northern end wasn't only remote but even dangerous. It was closed to the public for much of the year because of the risk of malaria, and the elephants themselves were thought of as a hazard. My well-worn copy of James Stevenson-Hamilton's *Our South African National Parks*, published in 1940, told me that you might see bull elephants alone or in small groups next to the road if you went up north, but

> [t]he large herds, containing females and young, are shy and keep away from the roads; which is fortunate; since the former might react dangerously did they suspect any threat of harm to their children.

By the end of the 1950s breeding herds *were* being spotted from the roads, but they were still thought of as dangerous. "Do not stop to photograph a herd of elephants with cows and calves," Eric Robins warned visitors to Kruger in his 1963 *Africa's Wildlife*, "but make haste to get away." I remember promising myself after my first trip to Kruger that, when I was a grown-up with my own car, I would travel to those mysterious and inaccessible northern regions. Perhaps I might even get a look (off in the distance, as my imagination saw it, through gaps between the green trunks of the fever trees) at the cow elephants and their young.

I remember, too, assuming that there had never *been* elephants in the southern part of the Kruger Park—that it just "wasn't elephant country." That was quite wrong. Elephants had been abundant there until ivory hunters equipped with firearms started operating in the area in the 1840s. The last elephants were seen there sometime in the early 1880s. By 1900 there were—apart from odd individuals wandering in from what was still "Moçambique"—very few elephants anywhere in the territory that would later become the Kruger Park.

In the early years of the twentieth century, conservationists managed to put an effective stop to hunting in that territory. As far as the elephants and rhinos were concerned, they were closing the stable door after the horse was gone. Elephants started filtering back, though, into the protected area from neighboring Mozambique as a result of the hunting ban. By the time South Africa's northeastern frontier area was legislatively proclaimed as the Kruger National Park in 1927, the northern end of that area contained about 100 of them. At the time of my first visit to Kruger in 1954, that 100 had grown to about 700, and a small number of elephants had, in fact, already re-entered the southern end of the Park. Few people had, however, heard about these returnees, let alone seen them.

* * *

Fast-forward to 1991. I'm in the Kruger Park again—this time with my ex-spouse and close friend Delphine du Toit, and our teenage son Nik. Branches are cracking like rifle shots as a herd of elephants feeds about fifty yards from the road. I'm watching a young female through my binoculars. She's prying a piece of bark off a fifteen-foot *Acacia nigrescens* tree with one of her tusks. Clamping a detached part of the bark between her enormous, six-pound molars, she backs up and pulls. The tree bends to near breaking before its foliage lashes violently upright as a sheet of bark rips free.

The 700 elephants that lived in Kruger at the time of my first visit nearly forty years before, have now become 10,000, and the animals have long since completed their reoccupation of the southern part of the Park. The southern end is, in fact, where I'm watching the young female stripping the *nigrescens*.

Delphine—our best animal spotter—says that there's a baby with the herd we're watching. I haven't seen it yet, but that's not surprising— the elephants are feeding in thick bush. Delphine has noticed, too, that the herd is moving along a path that will intersect with the road a few hundred yards behind us. She suggests that Nik turn around, drive back a quarter mile or so, do another U-turn, and wait for them.

While we're doing this, I'm struck again by how different the bushveld looks now that elephants have returned to the area. The southern Kruger Park was a neat, literally park-like place when I first saw it as a child, but now, since the elephants have returned, there are dead trees all around us, and many of the living ones are coppiced or growing at odd angles. Even trees that have been pushed into a prone position can survive, I've noticed, if some of their roots have kept a hold on the ground—the horizontal trunks send up vertical branches that eventually look like a row of new trees.

Delphine has guessed right about where the elephants are planning to cross the road—they emerge from the bush immediately ahead of us. We're excited to see that they've got not one, but *two* babies with them. A station wagon drives up and stops on the opposite side of where the animals are preparing to cross. The matriarch signals the other elephants to wait, and walks onto the road alone. Turning toward the station wagon, she spreads her ears and raises her trunk. The station wagon gives a little jerk as its driver shifts from park into reverse, but it only backs up a yard or two. That, I'm interested to see, is enough for the elephant—she seems more concerned with acknowledgment than with space. She swings her enormous body around and walks over to our car. Now it's Nik's turn to shift into reverse. Towering over us, the big cow extends her ears stiffly. We're all a bit nervous, and, with the car in reverse gear, Nik pushes the gas pedal too hard. The wheels spin in the dirt, showering pebbles at the elephant's front feet. For a second or two she looks down at us impassively through her long, crinkly eyelashes, then she turns away and signals the rest of the herd to follow her across the road.

The matriarch is the biggest of the ten or so elephants whose soft, ponderous steps are taking them between our car and the station wagon. Up close we see that the larger of the two babies is six months to a year old. The smaller one can't have been around for more than a few weeks. Its legs twinkle back and forth as it runs between the great, slow-moving limbs that surround it.

* * *

In a world in which more species are disappearing with each passing year, time runs backward in the Kruger National Park—not only elephants but Africa's two rhinoceros species have come back from near-extinction to form viable populations there.

Like humans and chimpanzees, white and black rhinos share a relatively recent common ancestor. Hominids and early white rhinos split, in fact, from their more forest-adapted relatives at roughly the same time in response to the appearance of relatively grassy conditions in the late Miocene/early-Pliocene of 6–8 million years ago. It's no exaggeration, therefore, to say that white rhinos and hominins have been part of each other's lives right from their origins.

The white rhinoceros is a giant that weighs, on average, about twice as much as the black rhinoceros. Perhaps because many people's ideas about African wildlife are shaped by the fauna of East Africa, the white rhino—which isn't found in Tanzania or Kenya—remains a relatively poorly-known animal. The drama of its restoration from near extinction to a state of relative abundance took place, moreover, in South Africa, which was, until the early 1990s, cut off from the rest of the world by the policies of its former government.

Fifteen thousand years ago white rhinoceroses *did* live in East Africa. Like the savannas themselves, with their characteristic acacia, combretum and commiphora scrub, the white rhino's range stretched unbroken between Southern Africa's Kalahari Desert and Northern Africa's Sahara. Then, starting around 11,000 years ago, Africa was abruptly subjected to a regime of enormously increased rainfall. The forests of the Congo Basin advanced into East Africa in response to it, cutting the grassy savannas of northern Africa off from those of the southern part of the continent and dividing the grass-eating white rhino population, in the process, into a northern and a southern group.

The pluvial or rainy regime gave way to a drier climate 4,000 years ago, allowing grassland to re-occupy what were once again becoming the Serengeti plains, but the northern and southern white rhino populations did not re-enter East Africa to rejoin each other. Along with a great many other dry-savanna species, ranging from the pygmy falcon *Polihierax* through the bat-eared fox to Kirk's dik-dik, white rhinoceroses remained divided into widely separated northern and southern pockets. Before humans wiped it out, the northern pocket of white rhinos was situated in an area where Chad, the Sudan, the Central African Republic, and the northern ends of Uganda and the DR Congo come together. The southern pocket lay some 1,200 miles away, in southern Africa south of the Zambezi River.

The southern pocket of white rhinos was the first to face the destructive power of firearms. Dutch colonists had established themselves at the southern tip of Africa as long ago as 1652, and, as they moved into the interior of Southern Africa during the eighteenth and nineteenth centuries, the white rhino was among the first animals to disappear from the areas they settled. Despite the fact that malaria and sleeping sickness delayed the intrusion of European settlers into the wilderness protected by the present-day Kruger Park, the area was, as I've mentioned, heavily hunted toward the end of the nineteenth century. The last white rhinoceros was seen there in 1894. By 1895 it was generally assumed that the entire southern population of white rhinos had ceased to exist. Because the existence of the northern pocket was not yet "known" at the time—that is, known to non-Africans—it looked as if humans had just exterminated the planet's third-biggest land animal.

It was then discovered—"to everyone's surprise," Ian Player tells us—that a handful of these giant rhinos had survived about 100 miles south of the present-day Kruger Park at the confluence of the Black Umfolozi and White Umfolozi Rivers in Zululand. A certain C. R. Varndell responded to this discovery by organizing a hunting expedition into the area. Varndell and his friends shot six white rhinos. Judging from what he wrote about it in his *Nature and Sport in South Africa*, published in 1897, H. A. Bryden seemed to think that this was a good idea:

> "There can, I fear, be little doubt that this rare and interesting quadruped will within the next two or three years have become quite exterminated—a creature of the past. Naturalists will have to thank... Mr. Coryndon and Mr. Varndell for their skill and success in procuring the first—and probably the last—complete specimens of this mammal before its final extermination."

Others reacted in a more rational way to this attack on the last population of giant rhinos. A conservationist by the name of C. D. Guise wrote to the Governor of Natal, Sir Walter Hely-Hutchinson, demanding that the animals be given protected status as "royal game." When Sir Charles Saunders, resident commissioner of Zululand, issued a proclamation to that effect in April of 1897, there were somewhere between forty and a hundred white rhinos left in the Umfolozi area. Even if an isolated animal or two still survived in the remote areas of what are now Botswana or Zimbabwe at that point, it was quite clear that this handful of Umfolozi survivors constituted the southern white rhinos' last viable population.

That population responded so well to protection that, by the start of the 1960s, there were almost 1,000 white rhinos in Zululand. The small reserves in that area that had been created for their protection were, however, filling up at this time, and it seemed that the remarkable resurgence of their population might start to level off. Not far away from them, however, the Kruger National Park—bigger than Tanzania's Serengeti National Park and Kenya's Masai Mara put together—still lay empty of rhinos. Beginning in 1961, therefore, something that conservationists had been urging since the 1920s was finally undertaken: white rhinos were captured in Zululand and released into Kruger.

* * *

The smaller, more secretive black rhino managed to hang on in the Kruger area longer than its giant white cousin. By the time hunting was prohibited there at the end of the nineteenth century, its population had, however, fallen to a critical low. Isolated individuals survived in the Kruger Park into the 1940s, but the species had, years before, declined to the point of no return. Beginning in 1971, therefore, black rhinos (which had survived in viable numbers in the Umfolozi area in the shadow of the protection given to their giant white cousins) were translocated to Kruger from Zululand. Other black rhinos were brought in from Zimbabwe. By 2020 Kruger was, as the result of these efforts, home to some 280 of them.

Returning the black rhinoceros to Kruger was originally seen as a matter of restoring the area's original fauna rather than a way of saving the species itself—there were, after all, still at least 50,000 black rhinos in East and Central Africa when that species was brought back to Kruger in 1971. During the next twenty years, however, the black rhino was eliminated from the east and central African savannas as rapidly and completely as the bison were wiped off the North American prairies a century earlier. Uganda, Kenya, Tanzania and Zambia lost well over 95 percent of their black rhinos during the 1970s and '80s. Even a famous sanctuary like the Serengeti National Park—still thought of by many people as an inviolable African wilderness—lost virtually all its rhinos. The few hundred black rhinos in Kruger had become one of the few viable populations of this species in Africa.

The rhino-killing spree of the 1970s and '80s had also, shortly after the turn of the twenty-first century, wiped out the northern pocket of white rhinos. Kruger had, by the year 2000, become one of the last places in Africa where Africa's two rhinoceros species were present in numbers large enough to ensure their continued viability. Unfortunately, however, this story of ecological resurrection must end on an uncertain note. As the value of rhino horn rose to the point where its value exceeded that of gold and cocaine, poaching started up in South Africa itself. Since 2008 the heartening population increases in both black and white rhino numbers in Kruger have given way to substantial declines in the numbers of both species (the white rhino population, now standing at about 3,500, is down from its 2008 level of 9,000). That would be a sad development in any circumstances, but the fact that it involves two species previously rescued from the brink of extinction adds to its poignancy.

* * *

Protecting the wilderness that sustains Kruger's megafauna involves the incidental protection of a huge number of smaller species that can be as fascinating as the Park's big mammals, reptiles and birds.

Watching a gathering of elephant families on the Mphongolo River in Kruger's northern end, I saw what looked like a hummingbird hovering around the whitish flowers of a *Combretum* bush next to the open window of my car. I did a double-take because there *are* no hummingbirds in Africa, and realized that I was looking at a large moth with a beak-like proboscis. Viewed from above, the hovering moth's color was a soft fawn, but its underside was marked with dazzling blacks and whites. My gaze went from the moth to a small, lemon-yellow spider crawling on the combretum bush and from there, to a busy ant-traffic moving along its branches. Two species of wasp were, I noticed, also visiting the bush—a black one so slender that I mistook it, at first, for a damselfly, and a larger bluish kind with a bright yellow abdomen.

My attention was drawn back to the elephants for a moment when a big bull caused excitement in one of the families by testing the sexual readiness of a cow with his trunk, but it returned to the bush next to my window when an enormous black wasp with red legs approached for a landing. Metallic blue beetles with yellow heads were also visiting the bush, and a blood-red fly hovering around its flowers caught my eye. It was impossible, though, to make out the details of all the flies, bees, moths and butterflies that were trying to keep station with those sparse, off-white flower clusters as they bobbed around in the wind.

I feel the same fascination that the combretum bush and its visitors produced in me, when I see *Barbus mariquensis* fishes flashing like bars of gold in the fast waters of Kruger's Letaba River, or when a blemish on a tile in the Nwashwitshaka research camp's washroom turns out, on closer inspection, to be a white tree-frog. It comes to me, too, when I stop to look at flowering knobthorn acacias and wild pear trees the way most people stop to look at buffaloes or lions.

Late winter/early spring is a great time to be in the Park. That's when the leafless branches of the *Erythrina* trees display their impossibly red flower clusters against the deep blue African sky, and when the gardenia trees are covered with white aromatic blossoms. My Palgrave's *Trees of Southern Africa* is open all the time. Maybe, I think, I'll discover some deeper truth about this remarkable place if I can identify *that* tree or *this* little bush. Biodiversity is long, though, and my botanical knowledge, short. I can't find the bush I'm looking at in Palgrave, or in Codd's *Trees and Shrubs of the Kruger National Park*. The only thing I can tell about it is that it has pods, and that it's a member, therefore, of the pea-family Leguminosae. Those pods are what drew my attention to it in the first place—how, I wonder, can they be *that* long and *that* thin? Evolution seems to have created Africa's Leguminosae the way Mozart wrote piano sonatas—always coming up with a more humorous, more unlikely, and more poignant variation on a theme as simple as peas in a pod.

I've been fascinated by the natural world for as long as I can remember. Growing up in Johannesburg, I spent a lot of my time on the banks of a tiny creek that wound its way through the rural suburb of Sandown. That creek, too small to have a name, flowed into a larger stream called Sandspruit, which ultimately joins the Limpopo River. Big freshwater crabs lived in it, and I discovered that I could catch them with string and a piece of meat. Expert handling was required to avoid claws that could—and did—draw blood. I was usually satisfied with a standoff, allowing the crabs to retreat backward into the water with raised claws after I'd pulled them up onto the bank.

It wasn't necessary to use a hook to catch the crabs—they simply hung on to the meat that I'd tied to the string. I noticed, though, that the meat would sometimes be grabbed by hard-to-see grayish creatures that *would* let go before I could pull them clear of the water. I solved the mystery of the gray meat-eaters by presenting the bait on a bent pin: they turned out to be "platannas" (*Xenopus laevis*)— frogs that were so exclusively aquatic that they kicked along on their bellies on land, unable to sit or hop.

Dipping a net made of a sheet of flyscreen into one of the creek's pools to catch tadpoles, I was amazed to see it come up with small silver fishes bouncing on its mesh. They were the first wild freshwater fishes I can remember seeing. I slipped some of them into a bottle to take home to my parents' fishpond. The thick glass near the bottom of that old-style Coke bottle magnified their heads enough for me to notice that they had two pairs of delicate, semi-transparent "barbels" hanging like mustache hairs from their upper lips. Those mustaches rang a bell. My grandfather, a zoologist-paleontologist whose work is discussed in Chapter 9, had given me a book on fish and fishing in India by a nineteenth-century colonial official called H. S. Thomas—and the position and relative size of my little fishes' barbels exactly matched those of a huge Indian freshwater fish called "mahseer" whose illustration I'd seen in that book.

Looking again at the illustrations in Thomas's book, I found that the resemblance between my little Johannesburg fishes and the great Indian mahseer was astonishingly close. It wasn't only visible in their barbels, but also in their relatively long and round bodies, the position of their eyes and mouths, and the shape of their fins. How, I wondered, could my little two-inch fishes be so similar to a fish that regularly reached four feet in length and had been known to reach ten?

The two species were similar enough, I suspected, to be closely related. I knew from Thomas's book that the mahseer was *Barbus tor*, so I guessed that my little fishes might also be *Barbus* something or other. That turned out to be right—when Rex Jubb's *Freshwater Fishes of Southern Africa* appeared some years later, I found out that they were *Barbus motebensis*.

Like many of the big nineteenth-century genera, *Barbus* has now been broken up, leaving the mahseer in the genus *Tor*, but that doesn't, of course, invalidate the accuracy of my guess about the close relationship between my little fishes and the Indian mahseer.

It struck me that my barbs and those that inhabited India must have a relatively recent common ancestor, and that the descendants of that ancestor must therefore have spread from Asia to Africa or vice versa. (Asia is, it happens, their ancestral home.) This gave me a fresh understanding of the fact that the natural world has a real and complex history. Making the connection between *Barbus motebensis* and *Barbus tor* made me realize, too, that I could unearth bits of that history. The pleasure of seeing those bits for myself was not diminished, moreover, by the fact that other people might have seen them before me.

Back in South Africa in the year 2000 after an absence of twenty-five years, I decide to see what the pools in which I caught my little barbs look like now. When I finally find my little creek, I see that townhouses have been built along its banks. The complex on the south bank, where I've parked my car, is called "Glenwood,"—a name as denatured and bland as the environment itself has become. "Glenwood" is surrounded by a tall, spear-pointed palisade fence bearing announcements that its security company provides a twenty-four-hour armed response. I'm told at the entrance that there's no access to the water, but I'm able to get through to my section of the creek, nonetheless, from the Linden Street Bridge with nothing but a small razor-wire scratch on one leg. I know I'm not going to find the ecosystem I experienced here in the '50s, but that still doesn't prepare me for the extent of the change.

The creek is still flowing, but the pools I knew as a child are hard to recognize because of the rubble and chunks of concrete that have been dumped into them. The ravine through which the creek runs is darker, too, than I remember it—the open veld with its wild carnations and amaryllises and its *Triandra* red-grass had been replaced by a gloomy weed-forest of willows and poplars from Europe, *Solanum* from South America, and eucalypts from Australia. With a start I recognize, however, the shapes of the reddish quartz outcrops that surround the pool in which I caught my *Barbus motebensis*.

The water in that pool has a nauseating smell now, and it's covered with bobbing flotillas of Styrofoam. The alien vegetation on its banks is festooned with the shreds of flood-borne plastic bags. Those shreds wave like sad pennants over drifts of metal, Styrofoam and plastic containers. I pick up a rusty aerosol can and make out the barely legible words "Keep out of reach of children."

Why am I braving this sewage smell to poke through this debris? Am I expecting to find a clue to the cause of the disaster? That's not going to happen—this maroon jar tells me only that Easy Waves Creme Relaxer has been specially formulated to relax coarse to resistant hair; this red plastic bottle, that Valvoline Automatic Transmission Fluid Type II D meets the stringent requirements of General Motors. I can't blame Estée Lauder or Colonel Sanders for what has happened here—it feels more like a tragedy than a crime.

A few months before I got back to South Africa, my father-in-law died at the age of eighty-nine. He'd remained an interesting and intellectually stimulating man to the end, and I miss him. While I was writing this chapter, my brother-in-law emailed from Cape Town to say that the family had received permission to place a memorial bench to his dad on Robberg, a nature reserve on South Africa's southern coast where the old man had fished and walked for years. I was asked for my thoughts about the wording of the memorial, but, toying with phrases like "In memory of Dr. ___ ___ who loved this peninsula," I realized that my mind was on another memorial—one I wanted to erect on the Linden Street bridge next to my little creek. It would say:

> In memory of *Barbus motebensis*, a small fish species that inhabited this stream up to the middle of the 1960s, when the water became too polluted to sustain its continued presence.

Notes and Sources

SECTION 1:

SHAKING THE FRAMEWORK OF THE GLOBE

CHAPTER 1: "ALL THE HUGEST, AND FIERCEST, AND STRANGEST FORMS HAVE RECENTLY DISAPPEARED"

bones of the elephant-sized ground-sloth: Charles Darwin, *Journal of Researches Into the Geology and Natural History of the Various Countries Visited by HMS Beagle*, "Ch VIII Banda Oriental and Patagonia" (London: John Murray 1845) Second edition p. 3.

reflect on the changed state of the American continent without the deepest astonishment: Charles Darwin, *Journal of Researches Into the Geology and Natural History of the Various Countries Visited by HMS Beagle* (1840) p. 173.

friend and mentor Charles Lyell: Lyell set out to prove that all geologic processes were due to natural, rather than supernatural, events. He was one of the first to propose these processes actually happened very slowly, and that the Earth was extremely ancient rather than a few thousand years old as most Bible scholars believed. In 1829, Lyell wrote his most famous work, *Principles of Geology*. Charles Darwin was greatly influenced by Lyell's ideas of a slow, natural change of geological formations. Captain Fitzroy gave Darwin a copy of *Principles of Geology*, which the latter studied during the *Beagle* voyage.

we must shake the entire framework of the globe: Charles Darwin, (Ed.) R. D. Keynes, *The Beagle Diary* (original 1834) Cambridge, Cambridge University Press (First edition 1988). The full quote: "The mind at first is irresistibly hurried into the belief of some great catastrophe: but thus to destroy animals, both large and small, in Southern Patagonia, in Brazil, on the Cordillera of Peru, in North America up to Behring's Straits, we must shake the entire framework of the globe. An examination, moreover, of the geology of La Plata and Patagonia, leads to the belief that all the features of the land result from slow and gradual changes."

we live in a zoologically impoverished world: James Marchant and Alfred Russel Wallace: Letters and Reminiscences, Vol. 1 (of 2), *The Project Gutenberg EBook*. Digital and Multimedia Center (Michigan State University Libraries 2005).

convinced that the rapidity of the extinction of so many large mammalia is actually due to man's agency: Alfred Wallace, *The World of Life: A Manifestation of Creative Power, Directive Mind and Ultimate Purpose* (New York: Moffat, Yard 1911).

"been 'proven' as thoroughly as any historical hypothesis can be": John Alroy, "Putting North America's end-Pleistocene megafaunal extinction in context: large scale analyses of spatial patterns, extinction rates, and size distributions." *Extinctions in Near Time: Causes, Contexts, and Consequences.* (Eds.) R. MacPhee and H. Sues (New York: Plenum Publishing 1999) pp. 105–143.

After they had been occupied by our species, the Mediterranean islands lost: Jacques Blondel, "On humans and wildlife in Mediterranean islands," *Journal of Biogeography,* Vol. 35, No. 3 (March 2008) pp. 377–575.

humans reached the Caribbean islands between 6,000 and 3,000 years BP: Matthew F. Napolitano, Robert J. DiNapoli, Jessica H. Stone, et al. "Reevaluating human colonization of the Caribbean using chronometric hygiene and Bayesian modeling," *Science Advances* Vol. 5, No. 12 (18 Dec 2019).

normal, giant, colossal, and titanic: Jared Diamond, *The Third Chimpanzee: The Evolution and Future of the Human Animal* (New York, Harper Collins, 1992) p. 328.

after Polynesian seafarers reached Hawaii: Storrs L. Olson and Helen F. James, "The role of Polynesians in the extinction of the avifauna of the Hawaiian Islands," *Quaternary extinctions: A prehistoric revolution,* (Eds.) P. S. Martin and R. G. Klein Tucson (University of Arizona Press 1984) pp. 768–780.

After reaching the world's fourth-largest island, Madagascar, at a time calculated to be somewhere between 920 CE and 670 CE: Atholl Anderson, Geoffrey Clark, et al., "New evidence of megafaunal bone damage indicates late colonization of Madagascar," *PLOS ONE,* Vol. 13 (10) (2018).

it took them only about a hundred years to exterminate the islands' avian megafauna: Morten Erik Allentoft, Rasmus Heller, et al., "Extinct New Zealand megafauna were not in decline before human colonization," *PNAS* Vol. 111 (13) (April 1, 2014) pp. 4922–4927.

the rats that had accompanied them multiplied hugely and killed off an enormous "minifauna": Ross D. E. MacPhee and Hans-Dieter Sues, *Extinctions in Near Time: Causes, Contexts, and Consequences* (Boston: Springer 1999).

In 1627, the last aurochs, the ancestor of domesticated cattle, was killed: "By the 13th century, the range of the aurochs had been significantly reduced as a result of hunting and habitat degradation. The last wild population survived in Poland's Forest of Jaktorów, where it was managed for centuries and protected by law. The population was carefully monitored until the late 16th century, when concern for its conservation began to fade...the aurochs was allowed to decline further, suffering from diseases transmitted by livestock and competing for resources with domestic cattle, until the last animal died in 1627." https://truenaturefoundation.org/aurochs.

hunters from Kamchatka wiped out *Hydrodamalis gigas*, an elephant-sized dugong: Lorelei D. Crerar, Andrew P. Crerar et al., "Rewriting the history of an extinction—was a population of Steller's sea cows (*Hydrodamalis gigas*) at St. Lawrence Island also driven to extinction?" *The Royal Society*—Palaeontology (2014).

extinctions of such conspicuous species included the passenger pigeon: Barry Yeoman, "Billions to None—Why the Passenger Pigeon Went Extinct," *Audubon* (May–June 2014).

According to the WWF's 2020 *Living Planet Report*: https://livingplanet.panda.org

literally hundreds of vertebrate species have followed it into extinction: Gerardo Ceballos et al., "Accelerated modern human-induced species losses: Entering the sixth mass extinction," *Science Advances* Vol. 1, no. 5 (19 Jun 2015).

Non-marine mollusks...are, for instance, being hit particularly hard: Charles Lydeard et al., "The Global Decline of Nonmarine Mollusks," *BioScience* Vol. 54, No. 4 (April 2004) pp. 321–330. "A staggering 42% of the 693 recorded extinctions of animal species since the year 1500 are mollusks (260 gastropods and 31 bivalves); this is more than the total (231) of all tetrapod species that have gone extinct during the same period. Nonmarine species constitute 99% of all molluscan extinctions. Although terrestrial vertebrate extinctions are well documented, invertebrate extinctions often go unnoticed by the general public, by most biologists, and by many conservation agencies. Only a tiny fraction (< 2%) of known molluscan species have had their conservation status properly assessed. Thus, the level of molluscan imperilment is poorly documented and is almost certainly underestimated," (page 322).

and it is expected to rise to at least a thousand times greater or more in the next few decades: E. O. Wilson, *The Creation: An Appeal to Save Life on Earth* (New York, W. W. Norton 2006).

CHAPTER 2: A UNIQUE AND DESTABILIZING CAPABILITY

a great white shark seized him by the thigh: Jonathan Kathrein and Margaret Kathrein, *Surviving the Shark: How a Brutal Great White Attack Turned a Surfer into a Dedicated Defender of Sharks* (New York: Skyhorse Publishing, 2012).

with all his noble qualities, with sympathy which feels for the most debased: Charles Darwin, *The Descent of Man, and Selection in Relation to Sex* (John Murray: London 1871).

the human family entered what they describe as "the cognitive niche": J. Tooby and I. DeVore, "The Reconstruction of Hominid Behavioral Evolution Through Strategic Modeling," *The Evolution of Human Behavior: Primate Models,* (Ed.) Warren G. Kinzey. Albany (NY: SUNY Press, 1987) p. 183.

A worldwide web of such genetic arms races linked each member: Ernst Häckel *Natürliche Schöpfungsgeschichte* (monograph), (Berlin: Reimer 1868) pp. 709 and 746.

...innovations that are too rapid with respect to evolutionary time: J. Tooby and L. Cosmides, "Unravelling the Enigma of Human Intelligence. Evolutionary Psychology and the Multimodular Mind," in *Evolutionary Intelligence*, (Eds.) Robert J. Sternberg, James C. Kaufman (New York: Psychology Press, 2013) p. 172.

The South Asian megafauna...was also partially protected from human-caused extinction by early exposure to our family: A. M. Jukarab, S. K. Lyons et al., "Palaeogeography, Palaeoclimatology, Palaeoecology," *Late Quaternary extinctions in the Indian Subcontinent*, Vol. 562 (Jan. 15, 2021), p. 110137.

The oldest evidence of hominin-made stone tools found so far: S. McPherron, Z. Alemseged, C. Marean, et al. "Evidence for stone-tool-assisted consumption of animal tissues before 3.39 million years ago at Dikika, Ethiopia," *Nature* 466 (2010) pp. 857–860.

a "sacred balance" with the natural world: David Suzuki, *The Sacred Balance: Rediscovering our Place in Nature* (Vancouver: Greystone Books, 2007).

Section 2:

LOST SERENGETIS

valuable information about the early history of our own species: Todd A. Surovell, Spencer R. Pelton, Richard Anderson-Sprecher, and Adam D. Myers, "Test of Martin's overkill hypothesis using radiocarbon dates on extinct megafauna": "...we argue that declines in the record of radiocarbon dates of extinct genera may be used as an independent means of detecting the first presence of humans in the New World. Our results, based on analyses of radiocarbon dates from Eastern Beringia, the contiguous United States, and South America, suggest north to south, time, and space transgressive declines in megafaunal populations as predicted by the overkill hypothesis." *PNAS* Vol. 113 (4) (January 26, 2016) p. 886.

CHAPTER 3: WERE GIANT TORTOISES THE FIRST VICTIMS OF HOMININ INGENUITY?

"Unlike so many people who get infatuated with their own theories": Mari N. Jensen, "Paul S. Martin, Pleistocene Extinctions Expert, Dies," *College of Science, University of Arizona News* (Sept. 16, 2010). Paul S. Martin developed the idea that early humans had hunted North America's Ice Age big game, including ground sloths, camels, mammoths and mastodonts, to extinction.

Africa's giant tortoises must have been even *more* vulnerable to growing hominin ingenuity: Wilhelm Schüle, *Prähistorischer Faunenschwund*, Op. cit. at pp. 132 et seq.

Lions...regularly try and fail to chew open the shells of tortoises of this size: Watch a Lion Try to Eat a Tortoise | Nat Geo Wild (https://www.youtube.com/watch?v=4NZ0JuDAyTo); Lion Tries to EAT tortoise! (https://www.youtube.com/watch?v=G-PogpX2z594).

the power of hominin inventiveness was, however, on an upward trajectory: S. McPherron, Z. Alemseged, C. Marean, et al., "Evidence for stone-tool-assisted consumption of animal tissues before 3.39 million years ago at Dikika, Ethiopia," *Nature* 466 (2010) pp. 857–860.

but their criticisms were muted five years later by the discovery, in 3.3-million-year-old deposits: Sonia and Lewis Harmand, Jason and Feibel, et al., "3.3-million-year-old stone tools from Lomekwi 3, West Turkana, Kenya," *Nature* 521 (2015) pp. 310–315.

a direct relationship between the extinction of giant tortoises in Africa with the appearance of early *Homo* and stone tool using behaviors: Terry Harrison, "Tortoises (Chelonii, Testudindae)," Chapter 17 in *Paleontology and Geology of Laetoli: Human Evolution in Context: Vol 2: Fossil Hominins and the Associated Fauna.* (Springer Science and Business Media, 2011) p. 498.

tortoise was a popular hominin menu item during the rest of the Pleistocene Epoch: M. C. Stiner, N. D. Munro, T. A. Surovell, "The tortoise and the hare. Small-game use, the broad-spectrum revolution, and paleolithic demography," Curr Anthropol. 2000 Feb, 41(1):39-79; see also David R. Braun, John W. K. Harris, et al., "Early hominin diet included diverse terrestrial and aquatic animals 1.95 Ma in East Turkana, Kenya," (Ed) Richard G. Klein, *PNAS* 2010 107 (22) (Stanford, CA: Stanford University 2010) pp. 10002–10007, as well as Teresa E. Steele, "A unique hominin menu dated to 1.95 million years ago," *PNAS* June 15, 2010 107 (24) pp. 10771–10772.

the disappearance of the giant tortoise species that had lived on several islands in the Mediterranean until around 8,000 years ago: Wilhelm Schüle, "Mammals, Vegetation and the Initial Human Settlement of the Mediterranean Islands: A Palae-oecological Approach," *Journal of Biogeography*, Vol. 20 (1993) pp. 399–411.

"I was always amused," Darwin wrote about the Galapagos giants: *Journal of Researches Into the Geology and Natural History of the Various Countries Visited by HMS* Beagle (1840) p. 384.

the same kind of intelligence test that Bernd Heinrich set his ravens: Bernd Heinrich, *Mind of the Raven: Investigations and Adventures with Wolf-Birds* (New York: Ecco 1999).

Chimpanzees have recently been observed breaking open tortoises' plastrons: Simone Pika et al., "Wild chimpanzees (*Pan troglodytes troglodytes*) exploit tortoises (*Kinixys erosa*) via percussive technology," *Scientific Reports* 9, Article number: 7661 (2019).

meat of the tortoises of those islands was "...so sweet, that no pullet eats more pleasantly": William Dampier, *A new voyage round the world.* Vol. 1 *[of Dampier's voyages.]* (James Knapton: Australia 1703).

tortoises tend to be much larger in the MSA [Middle Stone Age] layers than in the LSA [Later Stone Age] ones: Richard G. Klein and Kathryn Cruz-Uribe, "Middle and Later Stone Age large mammal and tortoise remains from Die Kelders Cave 1, Western Cape Province, South Africa," *Journal of Human Evolution* (2000), pp. 1 et seq.

Esmeralda, who lives on Bird Island: https://www.oldest.org/animals/tortoises.

factors such as reduced predation: Victoria L. Herridge and Adrian M. Lister, "Extreme insular dwarfism evolved in a mammoth," *Proc. R. Soc. B.* 279 (2012) pp. 3193–3200.

Makin v. Attorney General of New South Wales: ([1893] UKPC 56, [1894] AC 57 (12 December 1893), Privy Council (on appeal from NSW).

CHAPTER 4: STILL MARVELOUS BUT SIGNIFICANTLY REDUCED—THE MID-PLEISTOCENE DEPLETION OF AFRICA'S MEGAFAUNA

Dereck and Beverly Joubert: The Jouberts are environmentalists and documentary filmmakers who are National Geographic

Explorers-in-Residence. They work primarily in rural Botswana. They helped the National Geographic Society establish the Big Cats Initiative, which provides information and solutions to stop the number of big cats, especially lions, from dwindling.

a 1970 article in the *Zeitschrift für Säugertierkunde*: Vratislav Mazák, *Zeitschrift für Säugertierkunde* 35 (1970) pp. 359–362.

machairodont cats co-existed in Africa with the modern feline cats for a long period of time: Mauricio Anton, Àngel Galobart and Alan Turner, "Co-existence of scimitar-toothed cats, lions and hominins in the European Pleistocene. Implications of the post-cranial anatomy of *Homotherium latidens* (Owen) for comparative palaeoecology," *Quaternary Science Reviews* 24; 10–11; pp. 1287–1301 (2004) See also Alan and Anton Turner, et al., "Changing ideas about the evolution and functional morphology of Machairodontine felids," *Estudios Geológicos* (2011).

***Collins Guide to African Wildlife*:** Peter Alden, Richard Estes et al., *Collins Guide to African Wildlife* (New York: Harper Collins 2004).

the genus *Elephas* might have been represented, at that time, by as many as five coexisting species: N. E. Todd, "Trends in proboscidean diversity in the African Cenozoic," *Journal of Mammalian Evolution*, 13 (2006), pp. 1–10.

The elephants, hippos, and big perissodactyls we've just talked about: See generally *Collins Field Guide*. See also Theodore Haltenorth and Helmut Diller, *Mammals of Africa including Madagascar*, English translation. Collins (1984): pp. 13–26 for an excellent and extensive discussion of the African fauna that disappeared during the Pleistocene.

a much larger wave of extinctions hit Africa: M. E. Lewis and L. Werdelin, "Patterns of change in the Plio-Pleistocene carnivorans of eastern Africa," in *Hominin Environments in the East African Pliocene: An Assessment of the Faunal Evidence,* (Eds.) R. Bobe et al. (Springer, Dordrecht 2007) pp. 77–105; Lars Werdelin and Margaret E. Lewis, "Temporal Change in Functional Richness and Evenness in the Eastern African Plio-Pleistocene Carnivoran Guild," *2PLoS One.* 8(3). Published online (2013); Paul Martin, "Africa and Pleistocene Overkill," *Nature* 212 (1966) pp. 339–342.

The only region other than Africa to experience significant losses of big-animal species in the earlier part of the Pleistocene was South Asia: Julien Louys, Darren Curnoe, Haowen Tong, "Characteristics of Pleistocene megafauna extinctions in Southeast Asia," *PALAEO—Palaeogeography, Palaeoclimatology, Palaeoecology* 243 (2007) pp. 152–173.

in the vicinity of 2,000 to 1,500 years ago: "Humans may have colonized Madagascar later than previously thought", ScienceDaily, PLOS, www.sciencedaily.com/releases/2018/10/181010141739.htm (accessed April 3, 2021).

"Africa and Pleistocene Overkill": Paul Martin, "Africa and Pleistocene Overkill," *Nature* 212 (1966) pp. 339–342.

If Early Man was responsible for the destruction of the New World fauna, the argument goes: Paul Martin, "Africa and Pleistocene Overkill," Op. cit., at p. 339.

Paul didn't write anything further about the African Pleistocene: Paul Martin and Richard G. Klein (eds.), *Quaternary Extinctions: A Pre-historic Revolution* (Tucson, AZ: University of Arizona Press, Jan. 1, 1989) 3rd ed.

"The way I [Lars Werdelin] see it, did Climate Change Shape Human Evolution?": Symposium, Columbia University's Lamont-Doherty Earth Observatory (New York: Palisades April 19–20, 2012).

support for the idea that humans exterminated proboscideans in mid-Pleistocene Africa: Todd Surovell, Nichole Waguespack and P. Jeffrey Brantingham, "Global evidence for proboscidean overkill," *PNAS* (April 26, 2005), pp. 6231–6336.

his writings on the African disappearances: Wilhelm Schüle, "Prähistorischer Faunenschwund: Ursache und Wirkung des Absterbens," *Verlag Wissenschaft und Öffentlichkeit,* Freiburg im Breisgau (2001); "Human evolution, animal behavior, and quaternary extinctions": Schüle, "A paleo-ecology of hunting," *Homo* 41 (3) (1990) pp. 228–250; Schüle, "Landscapes and climate in prehistory: interaction of wildlife, man and fire," in J. G. Goldammer (Ed.) *Fire in tropical biota, ecosystem processes and global challenges.* Berlin, Heidelberg, New York: Springer Verlag (1990) pp. 273–318; Schüle (with Sabine Schuster), "Klima, Speer und Feuer. Zur ökologischen Rolle das Frühen Menschen," *Jahrbuch des Römisch-Germanischen Zentralmuseums*, Mainz 42 (1996) pp. 207–255.

Melville's *Moby Dick*, Schüle tells us, "ist ethologisch ein Unding": Wilhelm Schüle, *Prähistorischer Faunenschwund*, Op. cit., at p. 65.

lucky to survive the total destruction of his 4X4 truck by a bull elephant: George Wittemyer, "Firsthand account of vehicle attack by raging musth bull," www.savetheelephants.org/blog (2002) (Accessed Apr 6, 2021).

1990 *Homo* article I mentioned earlier: Schüle, "A paleo-ecology of hunting," Op. cit. (1990) pp. 228–250.

Hippopotamuses kill: Chris and Stuart Tilde, *Chris and Stuart Tilde's Field Guide to the Mammals of Southern Africa* (Fort Myers, FL: Ralph Curtis Publishing, 1989).

Pliocene and early-Pleistocene megaherbivores would not, Schüle reasoned in his 1990 "Paleo-ecology of Hunting"

article: Wilhelm Schüle, "A paleo-ecology of hunting," *Homo* 41 (3) (1990) pp. 228–250.

CHAPTER 5: EUROPE WITH SOME REMARKS ABOUT ASIA

Defecation is an even more complicated matter than urination in rhino society: Richard D. Estes, *The Safari Companion: A Guide to Watching African Mammals* (Hartford, VT: Chelsea Green Publishing, 1999) p. 191.

even extends to some of its surviving relatives in the odd-toed or "perissodactyl" order, horses and tapirs: Allen Keast, *Evolution, Mammals, and Southern Continents* (Chicago, IL: University of Chicago Press Journals, 1972).

"What…is the meaning of the six black dots?": Björn Kurtén, *The Ice Age* (London: Hart-Davis 1972) p. 125.

Mario Ruspoli says they are "a sign meaning the end": Mario Ruspoli, *Cave of Lascaux* (New York: Harry N. Abrams, 1987) (Reissue edition).

it seems clear that the Lascaux rhino: This statement is supported by the author's personal observation of male white rhinos in the Kruger National Park performing this ritual, as well as the many eyewitness video accounts documented on YouTube. Here are two vivid examples: https://youtu.be/MEmdNT1tEEw, https://youtu.be/p4xwRQSYlSk (Retrieved March 27, 2021).

top speed of big animals tends to be absolutely faster: R. McNeill Alexander, *Principles of Animal Locomotion* (Princeton, NJ: Princeton University Press, 2002) pp. 105–109.

megafauna included a member of the human family: Alex Menez, "The Gibraltar Skull: early history, 1848–1868," *Archives of* *natural history 45.1* (Edinburgh University Press, 2018) pp. 92–110 © The Society for the History of Natural History.

...die Herausbildung der Fähigkeit zum abstrakten denken, und das Vorhandensein einer Sprache: Dietrich Mania, Karl-heinz Fischer, Thomas Weber, *Bilzingsleben: Homo erectus, seine Kultur und seine Umwelt*, Volume 4 (Bilzingsleben, East Germany: Deutscher Verlag der Wissenschaften 1983).

CHAPTER 6: AUSTRALIA AND NORTH AMERICA

When the first humans arrived in Australia some 65,000 years ago: Chris Clarkson, Zenobia Jacobs et al., "Human occupation of northern Australia by 65,000 years ago," *Nature* 547, 76630 (20 July 2017) pp. 306–310. (On the fourth page of the article the authors state, however, that their approximately 65,000 estimate is conservatively reduced to 59.3 kyr, calculated as 65.0 kyr 204 minus the age uncertainty of 5.7 kyr at 95.4 percent probability.)

suggests that it might have been closer in size to a female lion or tiger: Stephen Wroe , Michael B Lowry and Mauricio Anton**.** "How to build a mammalian super-predator," *Zoology* (Jena) Vol. 111(3) (Mar 3, 2008) pp. 196–203.

Film footage of Thylacinus, taken in a zoo in the early thirties: 1936 footage, in captivity, https://youtu.be/bnT5vNE7LMI, https://youtu.be/bnT5vNE7LMI (Retrieved March 27, 2021).

Australian megafauna died out abruptly over a wide area: R. G. Roberts, Timothy F. Flannery, et al., "New Age for the Last Australian Megafauna: Continent-wide extinction about 46,000 years ago," *Science* Vol 292 (2001) pp. 1888–1892.

to establish that the youngest egg shell fragments of the giant *Genyornis* "ostrich": Gifford H. Miller and Julie Brigham-Grette,

"Amino acid geochronology: Resolution and precision in carbonate fossils," *Quaternary International*, Vol 1. Elsevier Ltd. (1989) pp. 111–128.

dominated by Araucaria trees were supplanted by plant communities dominated by ...eucalypts, which "rely" on fire: Bill Gammage, *The Biggest Estate on Earth* (Crows Nest, Australia: Allen and Unwin, 2011).

increase in burning, and an accompanying change in vegetation: C. S. M. Turney, "Development of a Robust 14C chronology for Lynch's Crater (North Queensland, Australia) using different pretreatment strategies," *RADIOCARBON*, Vol 43 (Tucson, AZ: University of Arizona 2001) pp. 45–54.

like a rash of ragwort on a bombsite: Colin Tudge, *The Time Before History: 5 Million Years of Human Impact* (New York: Scribner Book Company January 1, 1996) p. 206.

Recent research suggests that the American cheetah *Miracinonyx* may not be related to the Old World cheetah *Acinonyx*: Blaire Van Valkenburgh, Frederick Grady and Björn Kurtén, "The Plio-Pleistocene Cheetah-Like Cat Miracinonyx inexpectatus of North America," *Journal of Vertebrate Paleontology* Vol. 10, No. 4 (Abingdon, UK: Taylor & Francis, for the Society of Vertebrate Paleontology, Dec. 20, 1990) pp. 434–454.

the "horse" species abundantly represented in the Hagerman fossil beds in Idaho: Dr. H. Gregory McDonald "More Than Just Horses—Hagerman Fossil Beds," *Rocks & Minerals* 68:5 (1993) pp. 322–326.

the giant dugongs might have grown larger in the southerly portions of their range: Daryl P. Domning et al., "Steller's Sea Cow in the Aleutian Islands," *Marine Mammal Science* Society for Marine Mammalogy (2007).

these huge dugongs would probably have been as accessible to early human hunters: Georg Wilhelm Steller, *De Bestiis Marinis*, "The Beasts of the Sea" (St. Petersburg, Russia: St. Petersburg Imperial Academy of Sciences (1751).

CHAPTER 7: SOUTH AMERICA

probably able to kill horse-sized prey: E. O. Wilson, *The Diversity of Life*, Op. cit., p. 127.

"among the most mysterious members of the fauna of near time": [personal communication].

these big monkeys may have been at least partially terrestrial: Eckhard W. Heymann, Paul A. Garber, et al., *South American Primates: Comparative Perspectives in the Study of Behavior, Ecology, and Conservation* Springer Science and Business Media (2008).

the pioneers of rhino reintroduction weren't always conscious of the host of satellite creatures they were bringing back to Kruger along with the rhinos: Roan David Plotz, *The interspecific relationships of black rhinoceros (Diceros bicornis) in Hluhluwe-iMfolozi Park*, a thesis submitted to Victoria University of Wellington in fulfilment of the requirement for the degree of Doctor of Philosophy in Ecology and Biodiversity, p. 120.

Any of the big animals or birds that disappeared...could have been a "keystone" species: Andreas H. Schweiger and Jens—Christian Svenning, "Down—sizing of dung beetle assemblages over the last 53 000 years is consistent with a dominant effect of megafauna losses," *OIKOS Journal: Synthesizing Ecology* Nordic Society Oikos (2018).

In his *Diversity of Life* Edward Wilson explains, for instance, that when jaguars and pumas disappeared from Barro Colorado Island in Panama: E. O. Wilson, *Diversity of Life*, Op. cit., p. 165.

the microscopic animals feeding on the fungi and bacteria grow denser: Wilson, *Diversity of Life*, p. 168.

CHAPTER 8: LOOKING FOR OUR ANCESTORS

Darwin thought it probable that humans had evolved in Africa: Charles Darwin, *Descent of Man, and Selection in Relation to Sex* (London: John Murray 1871) pp. 198–199.

the living mammals are closely related to the extinct species of the same region: Darwin, *Descent of Man*, p. 199.

the ape-man who couldn't speak: Ernst Häckel, *Natürliche Schöpfungsgeschichte* (monograph) (Berlin: Reimer 1868) pp. 709 and 746.

human evolution might have been connected with Asia: Häckel, *Natürliche Schöpfungsgeschichte*, p. 746.

Initially Dubois couldn't make up his mind whether Darwin or Häckel was right: I'm indebted, for the account that follows of Dubois's discovery and thoughts, to L. T. Theunissen, *Eugène Dubois and the Ape-Man from Java: The History of the First "Missing Link" and Its Discoverer* (Springer Science and Business Media B.V. 1988), and to Pat Shipman's *The Man Who Found the Missing Link: Eugene Dubois and His Lifelong Quest to Prove Darwin Right* (Cambridge, MA: Harvard University Press, 2002).

Virchow, an anti-evolutionist: Jonathan Marks, "Why were the first anthropologists creationists?" *Evolutionary Anthropology* (11 January 2011).

Eugène Dubois and the Ape Man from Java: L. T. Theunissen, *Eugène Dubois and the Ape-Man from Java: The History of the First "Missing Link" and Its Discoverer* (Berlin/Heidelberg, Germany: Springer Science and Business Media B.V. 1988).

two more skulls that also seemed similar to Dubois's *Pithecanthropus*: Robin Dennell, *The Palaeolithic Settlement of Asia.* Cambridge World Archaeology (Cambridge, UK: Cambridge University Press 2008) p. 155.

related to each other in the same way as two different races of present mankind: Ian Tattersall, *The Fossil Trail: How We Know What We Think We Know about Human Evolution* (London: Oxford University Press 1995) p. 66.

of which erectness of carriage is one of the incidental characteristics: Grafton Elliot Smith, *Essays on the Evolution of Man* (London: Oxford University Press 1924) p. 39.

He [Elliot Smith]...rightly foresaw that before the anthropoid [i.e., ape-like] characters would disappear: Sir Arthur Keith, *The Antiquity of Man* (London, UK: Williams and Norgate 1915) p. 434.

a big, robust lower jaw that lacked the kind of protruding chin that *Homo sapiens* possesses: See generally Katerina Harvati, "100 years of *Homo heidelbergensis*—life and times of a controversial taxon," *Mitteilungen der Gesellschaft für Urgeschichte* 16 (2007) pp. 85–94.

Human skull fragments were found together with an ape-like jaw fragment in a gravel pit: My account of the Piltdown fraud is based on John E. Walsh, *Unraveling Piltdown: The Science Fraud of the Century and Its Solution,* Random House (1996), and on an eight-year-long, multidisciplinary study which included palaeobiologists, historians, dental experts and ancient DNA specialists: De

Groote, Flink, Abbas, et al., *New genetic and morphological evidence suggests a single hoaxer created 'Piltdown man'* (London, UK: Royal Society Open Science 2016).

Murray's rejection of the book...was my bitterest disappointment: Arthur Keith, *An Autobiography* (New York: Philosophical Library 1950) (First ed.) pp. 233–234.

"He has gone up," Oxford Professor of Geology, W. J. Sollas, wrote in 1924, "like a rocket and will come down like the stick": Quoted by Donald C. Johanson and Maitland Armstrong Edey, *Lucy, the Beginnings of Humankind* (Simon and Schuster 1981) p. 46.

Piltdown was, of course, exposed as a fraud in 1953: John Walsh, *Unraveling Piltdown: The Science Fraud of the Century and Its Solution* (Random House 1996).

the general thickness," he wrote, announcing the "discovery" of one of the skull fragments to Smith Woodward, "seems to me to correspond with the right parietal of *Eoanthropus*: Walsh, *Unraveling Piltdown*, p. 57.

Woodward's wife later vividly recalled this uneasy and troubled interlude: Walsh, *Unraveling Piltdown*, pp. 58–59.

over the next several months and years, accelerating with the war's end: Walsh, *Unraveling Piltdown*, p. 60.

betrays the forger's utter confidence in being able to manipulate: Walsh, *Unraveling Piltdown*, p. 196.

a harmless joke that developed into an embarrassment for its perpetrators: Stephen Jay Gould, in "The Piltdown Conspiracy," Chapter 16, in his essay collection *Hen's Teeth and Horse's Toes: Further Reflections in Natural History* (New York: W. W. Norton, 1994).

Shattock excluded the possibility of acromegaly and six other potentially bone-deforming conditions: S. G. Shattock, "Morbid thickening of the calvaria, and the reconstruction of bone once abnormal: A pathological basis for the study of the thickening observed in certain Pleistocene crania," *XXVIIth International Congress of Medicine Section VII.* (Published in London July, 1913) pp. 44–45.

The bone is naturally formed: Op. cit., Keith, *The Antiquity of Man,* p. 320.

So compelling was the honesty of Dawson's manner of speech: Letter written by Keith to Ashley Montagu on September 19, 1954, reproduced in Phillip V. Tobias, "An Appraisal of the Case Against Sir Arthur Keith," *Current Anthropology* (1992) p. 243.

CHAPTER 9: EX AFRICA SEMPER ALIQUID NOVI

I fell under his spell that night: R. A. Dart, *Adventures with the missing link* (London: Hamish Hamilton 1959) p. 27.

that even today I often find myself guided by the standards which he implanted in my young mind: Dart, *Adventures,* p. 28.

Working under Elliot Smith was my student dream come true: Dart, *Adventures,* p. 29.

He created an embryonic library: M. Dayal et al., "The History and Composition of the Raymond A. Dart Collection of Human Skeletons at the University of the Witwatersrand, Johannesburg, South Africa," *American Journal of Physical Anthropology* Vol. 140 (2009) pp. 324–335. See generally Dean Falk, *The Fossil Chronicles: How Two Controversial Discoveries Changed Our View of Human Evolution* (Berkeley, CA: University of California Press 2011).

Hans Reck, a German vulcanologist, paleontologist and adventurer, had found baboon fossils at Olduvai Gorge: Michael A. Cremo, *Searching for the Truth with...The Forbidden Archaeologist: The Atlantis Rising Magazine Columns of Michael Cremo* (Los Angeles, CA: Bhaktivedanta Book Publishing Inc 2010) p. 3.

I was careering down the hill in my Model-T Ford to discuss the skull and Taung with a friend and colleague: Cremo, "Searching for the truth," p. 3.

***three times as great* as in any existing endocast of a living ape's skull:** Op. cit., Dart, *Adventures,* p. 6.

"After considerable experience," Charles Darwin admitted in his treatise on barnacles: Charles Darwin, *The Works of Charles Darwin, Vol. 12: A Monograph of the Sub-Class Cirripedia, Vol. II: The Balanidae (Part One)* (New York: NYU Press 1989) p. 135.

implicit in the globular form of the skull which was obviously balanced on a more vertically placed type of backbone: Raymond Dart, *Adventures with the Missing Link* (New York: Harper 1959) p. 14.

he is one of, at the most, three or four men in the world who have the experience of investigating such material and appreciating its real meaning: Op. cit., Falk, *Fossil Chronicles,* p. 38.

he may just have identified "one of the most significant finds made in the history of anthropology": C. K. Brain, "Raymond Dart and our African origins" from Chapter 1 of *A Century of Nature: Twenty-One Discoveries that Changed Science and the World.* Laura Garwin and Tim Lincoln (Eds.) (Chicago and London: University of Chicago Press 2003).

He didn't claim that the Taung child was a member of the human family: Raymond Dart, "Australopithecus africanus The Man-Ape of South Africa," *Nature* Vol. 115 (1925) pp. 195–199.

in the same group or sub-family as the chimpanzee or gorilla: Sir Arthur Keith, "The Fossil Anthropoid Ape from Taungs," *Nature* Vol. 115 (1925) pp. 234–236.

Raymond Dart, "The Taungs Skull," *Nature* 116, No. 462 (1925) p. 11.

not unknown in the young of the giant anthropoids: Op. cit., Keith, "Fossil Anthropoid Ape," p. 235.

...chosen so barbarous a (Latin-Greek) name for it as *Australopithecus*: Keith, "Fossil Anthropoid Ape," p. 236.

***Australopithecus* is an unpleasing hybrid:** F. A. Bather, "The Word 'Australopithecus' and Others," *Nature* Vol. 115 (1925) p. 469.

"If you want to join in a game," a British Museum scientist named F. A. Bather lectured Dart in *Nature*, "you must first learn the rules": Op. cit., Bather, "The Word 'Australopithecus,'" p. 947.

reacted to the Wembley exhibit by announcing that Dart's claims were "preposterous": A. Keith, "The Taungs Skull," *Nature* Vol. 116 (1925) p. 11.

the unsurprising conclusion that it could not represent a human ancestor: A. Keith, *New Discoveries Relating to the Antiquity of Man* (Williams and Norgate 1931).

that Professor Dart was right and I was wrong: Op. cit., Dart, *Adventures*, p. 88.

They possessed to a degree unappreciated by living anthropoids the use of their hands and ears: Raymond Dart, "Australopithecus africanus The Man-Ape of South Africa," *Nature* 115 (1925) pp. 195–199.

a total absence of anything that modern humans would recognize as language: Op. cit., Leakey, *Origins of Humankind*, p. 125. N. Chomsky, in *Powers and Prospects. Reflections on human nature and the social order* (London: Pluto Press, 1996) p. 30, also argues that language appeared, like Athena, fully grown and fully armored.

and suddenly in a "great leap forward" is now being abandoned: Allison Brooks and Sally McBrearty, "The revolution that wasn't: a new interpretation of the origin of modern human behavior," *Journal of Human Evolution.* Elsevier (2000) p. 39.

its roots may well go down to our Australopithecine forebears: Steven Pinker, *The Language Instinct* (New York: Harper Perennial Modern Classics 2007) pp. 354–356.

The last of these conclusions was clearly wrong—no corroborating evidence: Richard Currier, *Unbound: How Eight Technologies Made Us Human, Transformed Society, and Brought Our World to the Brink* (Arcade, Reprint edition 2015) p. 58.

***Australopithecus* lived...a grim life:** Op. cit., Dart, *Adventures*, p. 189.

The loathsome cruelty of mankind to man is the inescapable product: Dart, *Adventures*, p. 198.

Not in innocence, and not in Asia, was mankind born: Robert Ardrey, *African Genesis: A Personal Investigation into the Animal Origins and Nature of Man* (Fontana Books, first published 1961) Opening sentence.

succeeded, probably, by being at times extremely unpleasant: Colin Tudge, *Time Before History* (New York: Touchstone 1997) p. 197.

led them as domesticated pets to his slaughterhouses: Op. cit., Dart, *Adventures*, p. 198.

But after this initial dismay, he became increasingly enthusiastic: Bob Brain, "Were our early ancestors murderers and head-hunters? A prehistoric detective story," *Quest* Vol. 5 (2) (2009) p. 18. A joint project of the National Research Foundation and the Department of Science and Technology, Republic of South Africa.

In 2010, two bones dated to 3.4 million years ago...with cut-marks probably made with stone tools: S. McPherron, Z. Alemseged, C. Marean, et al., "Evidence for stone-tool-assisted consumption of animal tissues before 3.39 million years ago at Dikika, Ethiopia," *Nature* 466 (2010) pp. 857–860.

deliberately manufactured stone tools showed up in 3.3-million-years-ago deposits: S. Harmand et al., "3.3-million-year-old stone tools from Lomekwi 3, West Turkana, Kenya," *Nature* 521 (2015) pp. 310–315.

"Phillip, I have to do it this way...": Martin Meredith, *Born in Africa: The Quest for the Origins of Human Life* (New York: Perseus Group, 2011) p. 201.

Dreyer found pieces of the face: Phillip V. Tobias, *Yearbook of Physical Anthropology, Vol 28*, History of Physical Anthropology in Southern Africa (1985) p. 16.

National Museum: Bloemfontein, South Africa: a natural history, cultural and art museum, established in 1877.

feeling nervous about the fact that Dreyer was holding the precious Florisbad cranium in an unsteady grasp: Personal communication.

Those tests revealed that the human whose remains Dreyer had unearthed had lived some 259,000 years BP: R. Grün, J. Brink, N. Spooner, et al., "Direct dating of Florisbad hominid," *Nature* 382 (1996) pp. 500–501.

its last-surviving and most remarkable member, *Homo sapiens*, also had its origin in Africa: "The potential of ESR to match the ability of AMS radiocarbon in directly dating human fossils is at last being realized. In 1996 the first of an increasing number of applications of this technique to significant human fossils was made when Rainer Grün and I collaborated with colleagues, including James Brink from South Africa, to date the Florisbad human skull. This fossil, which had been found in 1932, is actually rather incomplete but seems to combine a large and fairly modern-looking face with a strong brow ridge and somewhat receding forehead. For many years it was assumed to date from about 40,000 years ago, based on a radiocarbon date from peat deposits at the site, and on that basis it seemed to be a relic hanging on in the margins of southern Africa, while moderns were evolving and spreading in western Asia and Europe. As such, it supposedly demonstrated the backward role of Africa in modern human evolution—the primitive Florisbad humans merely marking time until moderns arrived from farther north and replaced them. However, the fossil preserved one upper molar tooth, and a tiny fragment of its enamel was taken to Rainer's lab in Australia for ESR dating—with sensational results. The fossil was not 40,000 but about 260,000 years old! Thus its potential role in human evolution was revolutionized in a stroke: rather than representing a southern African equivalent of the Neanderthals, on the brink of extinction, it could instead have been an ancestor to us all." From Chris Stringer, *Lone Survivors: How We Came to Be the Only Humans on Earth* (New York: Macmillan, 2012) pp. 47–48.

confirmed by the discovery, elsewhere on that continent, of several other hominin specimens closely resembling Dreyer's *Homo helmii*: Tim D. White, B. Asfaw, et al., "Pleistocene *Homo* *sapiens* from Middle Awash, Ethiopia," *Nature* 423 (2003) pp. 742–747, and Jean-Jacques Hublin et al., "New fossils from Jebel Irhoud, Morocco and the pan-African origin of *Homo sapiens,*" *Nature* 546 (2017) pp. 289–292.

CHAPTER 10: TRAVELING THROUGH TIME IN A CLUSTER OF SPECIES

tragelephine or "bushbuck-type" antelopes, and reduncine or "waterbuck-type" antelopes: Kristin Kovarovic, *Bovids as Palae-oenvironmental Indicators,* Thesis (London: University College London, 2004).

During the late Pliocene and early Pleistocene, that predilection for primates extended to australopithecines: see generally C. K. (Bob) Brain, *The Hunters or the Hunted? An Introduction to African Cave Taphonomy* (Chicago: University of Chicago Press, 1981).

Between 1932 and 1947, lions killed 1,500 people in the Njombe District in southern Tanzania: National Geographic Resource Library | VIDEO "Tanzania Terror—One lion sets a trap to catch his human prey (Sept. 21, 2011). (https://www.nation-algeographic.org/media/tanzania-terror) (Retrieved March 21, 2021).

A more recent report tells us that, since 1990, over 600 people had been killed by lions in that country: Lion attacks on human/Man-eating lions/Tanzania National Geographic Documentary (April 21, 2020) (https://www.youtube.com/watch?v=B6wKgE-NYLgA) (Retrieved March 21, 2021).

a slightly divergent toe and less well-developed arches in the Laetoli 3.6-million-year-old *Australopithecus afarensis* footprints and in the *A. africanus* bones of Stw 542 from Sterkfon-

tein: Carol Ward, "Bipedal Foot Morphology," *Centre for Academic Research and Training/Anthropogeny* (https://carta.anthropogeny.org) (Retrieved March 21, 2021).

bipedal performance at the expense of arboreal agility: The fourth metatarsal is useful in distinguishing between the feet of tree climbers and land walkers. Lucy's fourth metatarsal places her as a non-climber. C. V. Ward, W. H. Kimbel, and D. C. Johanson, "Complete Fourth Metatarsal and Arches in the Foot of *Australopithecus afarensis,*" *Science* 331 (2011) pp. 750–753.

Gorillas can climb trees...but they do so "with great caution": Jean Dorst and Pierre Dandelot, *A Field Guide to the Larger Mammals of Africa Hardcover* (London: Collins 1978) 2nd ed. p. 86.

"climb trees with consummate agility, and may occasionally jump to a nearby or lower tree...": Op. cit., Dorst, p. 89.

the molar teeth of the australopithecines were broader than those of apes and covered: Mark F. Teaford and Peter S. Ungar, "Diet and the evolution of the earliest human ancestors," *PNAS* Vol. s97 (25) (2000) pp. 13506–13511.

Looked...very different from each other as living beings, but have a very similar bone structure: R. E. F. Leakey and Alan C. Walker, "Australopithecus, Homo erectus and the single species hypothesis," *Nature* Vol. 261 (1976) pp. 572–574.

a 3.3- to 3-million-year-old jaw fragment: Michel Brunet, A. Beauvilain et al., "The first australopithecine 2,500 kilometres west of the Rift Valley (Chad)," *Nature* 16:378 (1995) pp. 273–275.

regarded as being distinct from _afarensis_: Meave Leakey, F. Spoor, et al., "New hominin genus from eastern Africa shows diverse middle Pliocene lineages," *Nature* 410 (2001) pp. 433–440.

partial tibia that includes part of the knee-joint: Meave Leakey, C. S. Feibel et al., "New specimens and confirmation of an early age for *Australopithecus anamensis,*" *Nature* 393 (1998) pp. 62–66.

the British-South African paleontologist Ron Clark found hominin foot bones in a museum storage box: Jamie Shreeve, "'Little Foot' Fossil Skeleton Rivals Famous Lucy in Age," *National Geographic* (Apr 1, 2015).

pushed the history of our family over the 5.3-million-year-ago boundary: Tim White et al., "A New Kind of Ancestor: Ardipithecus Unveiled, Published by AAAS," *Newsfocus, Science* Vol. 326 (2 Oct 2009) pp. 8–11.

may also have been intended to evoke the French word *aurore*: Martin Pickford and Brigitte Senut, "The geological and faunal context of Late Miocene hominid remains from Lukeino, Kenya," *Comptes Rendus de l'Académie des Sciences—Series IIA—Earth and Planetary Science* Vol. 332 (2001) pp. 145–152.

when hominins were quadrupedal: John Whitfield, "Oldest member of human family found," *Nature—News Feature* (Basingstoke, UK: Springer Nature 2002).

The walls of OH 7's skull were also somewhat thinner: L. S. B. Leakey, P. V. Tobias, J. R. Napier, "A new species of the genus Homo from Olduvai Gorge, Tanzania," *Nature* 202 (1964) pp. 308–312.

"a storm of objections": Richard Leakey, *The Origins of Human-kind* (New York: Basic Books 2008) p. 156.

direct evidence of deliberately manufactured stone tools found in 3.3-million-years-ago deposits: Sonia Harmand, Jason E. Lewis et al., "3.3-million-year-old stone tools from Lomekwi 3, West Turkana, Kenya," *Nature* 521 (2015) pp. 310–315.

"align more closely with *Homo* than with any other hominid genus": Brian Villmoare, William H. Kimbel et al., "Early *Homo* at 2.8 Ma from Ledi-Geraru, Afar, Ethiopia," *Science* Vol. 347, No. 6228 (20 Mar 2015) p. 1352.

are wide enough to blur the distinction between gracile and robust australopithecines: Berhane Asfaw, Tim White, et al. "*Australopithecus garhi*: A New Species of Early Hominid from Ethiopia," *Science* Vol. 284, No. 5414 (1999) pp. 629–635.

were probably made in the process of removing the animal's tongue: Jean de Heinzelin, J. Desmond Clark, Tim White, et al., "Environment and Behavior of 2.5-Million-Year-Old Bouri Hominids," *Science* Vol. 284, No. 5414 (1999) pp. 625–629.

In 1984 , the remarkably complete skeleton of a tall, long-limbed hominin boy dated to 1.5–1.6 million years ago: F. Brown, J. Harris , R. Leakey, A. Walker "Early *Homo erectus* skeleton from west Lake Turkana, Kenya," *Nature* Vol. 316, No. 6031 (1985) pp. 788–792.

Skull fragments attributed to the same species had been discovered at Swartkrans in South Africa in 1969: Donald Johanson and Blake Edgar, *From Lucy to Language* (New York: Simon & Schuster 1996) p. 184.

Valerii Alexeev, labeled it *Pithecanthropus rudolfensis*: Alexeev, https://humanorigins.si.edu/evidence/human-fossils/species/homo-rudolfensis (retrieved April 6, 2021) *Smithsonian National Museum of Natural History: Human Evolution Evidence: Homo rudolfensis.*

Strong brain lateralization is a uniquely human characteristic: Ralph L. Holloway, "The Evolution of the Hominid Brain," *Handbook of Paleoanthropology* (Berlin, Heidelberg: Springer 2015).

stone flakes belonging to the Oldowan Industrial Complex were, in the great majority of cases, struck off their "cores" by right-handed individuals: Nicholas Toth, "Archaeological evidence for preferential right-handedness in the lower and middle Pleistocene, and its possible implications," *Journal of Human Evolution* Vol. 14, No. 6, Elsevier (1985) pp. 607–614.

CHAPTER 11: HOMINIDS AS MEAT-EATERS

Utilizes carcasses of large vertebrates more efficiently than other carnivores: Richard D. Estes, The Safari Companion: A Guide to Watching African Mammals. Chelsea Green Publishing (1999) p. 291.

triggered human evolution, and propelled man to the creature he is today: Richard Leakey, *The Origins of Humankind* (New York: Basic Books 2008) at p. 62 Leakey cites Perper and Schrire from a conference paper dated 1963 (original currently inaccessible—Apr 6, 2021).

it was thought that she'd witnessed an aberrant phenomenon: "Jane Goodall Reporting from Gombe Stream Game Reserve, 8 Oct 1962: 'Dear Mr. Payne...Twice again during the month I have seen them with meat...once the prey was almost certainly a red colobus monkey...'" *National Geographic Resource Library: Article* (1962).

is now known to kill and eat flying squirrels, small forest antelopes and monkeys: Martin Surbeck and Gottfried Hohmann, "Primate hunting by bonobos at LuiKotale, Salonga National Park," *ScienceDirect* Vol. 18 (2008) pp. 906–907.

political allies and/or potential sexual partners: Craig B. Stanford, *Chimpanzee and Red Colobus: The Ecology of Predator and Prey* (Cambridge, MA: Harvard University Press, 1998).

Christophe Boesch has observed an equally high degree of cooperation: Christophe Boesch and Hedwige Boesch, "Hunting behavior of wild chimpanzees in the Taï National Park," *American Journal of Physical Anthropology* Vol. 78 (1989) pp. 547–573.

chimpanzees were reliably reported to have killed at least two human infants in that area for food: "To protect [her son] from the chimpanzees, Goodall built a protective 'cage' for him and never left his side as she did her research. The 'cage' was like a giant cot where Grub could crawl, stand and walk within it. He was never on his own or left alone until he was 3 years old." https://www.mothersinscience.com/trailblazers/jane-goodall (retrieved Apr 6, 2021).

"...[C]himpanzees...may be among the most important predators of certain prey species in the African ecosystems in which they live": Craig B. Stanford, "Wild Chimpanzees: Implications for the Evolutionary Ecology of Pliocene Hominids," *American Anthropologist* New Series, Vol. 98, No. 1.

The chimp-human line had, in other words, split off from the ancestors of gorillas: Ajit Varki and David Loren Nelson, "Genomic Comparisons of Humans and Chimpanzees," *Annual Review of Anthropology* 36 (2007)s p. 294.

the tool-scratched lower jaw of a "medium-sized alcelaphine antelope": Thomas Plummer, "Flaked Stones and Old Bones: Biological and Cultural Evolution at the Dawn of Technology," *American Journal of Physical Anthropology* Vol. 125 (2004) p. 118.

convinced that his little australopithecines could kill: Raymond Arthur Dart, "The predatory transition from ape to man," *International Anthropological and Linguistic Review,* Vol. 1, No. 4 (Leiden, Netherlands: Brill 1953).

another blow against the idea that early members of the human family could have hunted: Lewis Binford, *Ancient Bones* *and Modern Myths* (Amsterdam, Netherlands: Academic Press, Elsevier Inc., 1981).

cuts that could only have been made by hominins using stone tools to slice through the ligaments: Henry Bunn et al. "FxJj50: An Early Pleistocene Site in Northern Kenya," *World Archaeology* Vol. 12, No. 2 (1980) pp. 109–136.

Did the discovery of these cut marks turn things around: R. Potts and P. Shipman, "Cutmarks made by stone tools on bones from Olduvai Gorge, Tanzania," *Nature* 291 (1981) pp. 577–580.

"profoundly unfashionable" to talk about big-game hunting, or indeed *any* kind of hunting, by early humans: R. Dennell, "The world's oldest spears," *Nature* 385 (1997) pp. 767–768.

early human hunting of sizable mammals has largely been discarded: Ian Tattersall, *Becoming Human: Evolution and Human Uniqueness* (Boston: Houghton Mifflin Harcourt 1999).

rejection of hunting in early *Homo* has been too assiduous: Richard Leakey, The Making of Mankind (Dutton 1981) p. 73.

we'd resisted the idea that hominins had scavenged to feed themselves: Donna Hart, Robert Wald Sussman, et al., *Man the Hunted: Primates, Predators, and Human Evolution* (Boulder: Westview Press 2005).

"Females not only dominate males at kills and other favored sites, but also lead clan members on pack hunts, boundary patrols and into battle": Op. cit., Estes (1991) p. 292.

"Carnivores scavenge when they can," Pat Shipman tells us, "and hunt when they must": quoted Op. cit., Leakey (1981) p. 72.

the tool-marked bones found in association with some Neanderthal living areas could have been obtained from "older individuals": Ian Tattersall, *Becoming Human: Evolution and Human Uniqueness* (Boston: Houghton Mifflin Harcourt 1999).

These sites have also yielded the largest sample of cut-marked bones known from the time interval 2.6–2.3 million years ago: Manuel Dominguez-Rodrigo, and Travis Pickering, "Early Hominid Hunting and Scavenging: A Zooarcheological Review," *Evolutionary Anthropology: Issues, News, and Reviews* 12 (2003). See also Dominguez-Rodrigo, Pickering et al., "Cutmarked bones from Pliocene archaeological sites at Gona, Afar, Ethiopia: Implications for the function of the world's oldest stone tools," *Journal of Human Evolution* Vol. 48, No. 2 (2005).

"The major, or in many cases, the only usable or edible parts consisted," Binford wrote, "of bone marrow": quoted in K. Paddayya, "The Epistemology of Archaeology: A Postscript to the New Archaeology," *Bulletin of the Deccan College Research Institute,* Vol. 45 (1986) p. 96.

This giant hyena lived throughout Africa up until about 1.4 million years ago: Alan Turner and Mauricio Anton, *The giant hyaena, Pachycrocuta brevirostris (Mammalia, Carnivora, Hyaenidae)* (Amsterdam, Netherlands: Elsevier—Geobios 29 1996) pp. 455–468.

and keep chipping, and knock thirty flakes off a single core: Harmand, Lewis et al., "3.3-million-year-old stone tools from Lomekwi 3, West Turkana, Kenya," *Nature* 521 (2015) pp. 310–315.

a great deal more than just banging two rocks together: Brenda Fowler, "SCIENTISTS AT WORK: Kathy Schick and Nicholas Toth; Recreating Stone Tools to Learn Makers' Ways," *The New York Times* (Dec. 20, 1994).

Nine of the fifty-four implements examined showed wear traces: L. Keeley and N. Toth, "Microwear polishes on early stone tools from Koobi Fora, Kenya," *Nature* 293 (1981) pp. 464–465.

before **the advent of stone cutting tools:** Nancy Tanner and Adrienne Zihlman, "Women in Evolution," *Part I: Innovation and Selection in Human Origins* Vol. 1, No. 3. (Chicago: University of Chicago Press, 1976) pp. 585–608.

"Darwin was arguing…that the evolution of our unusual mode of locomotion was directly linked to the manufacture of stone weapons": Op. cit., Leakey (1981) p. 3.

and sharpen crude spears with their teeth to skewer bush babies hiding in hollow trees: Ann Gibbons, "Spear-Wielding Chimps Seen Hunting Bush Babies," *Science* 315 (2007) p. 1063.

The fossils give a satisfyingly clear and decisive answer: Richard Dawkins, *A Devil's Chaplain: Reflections on Hope, Lies, Science, and Love* (Boston: Houghton Mifflin 2013).

CHAPTER 12: BODIES, BRAINS, AND TECHNOLOGY

"tall, athletic and powerfully muscled": Richard Leakey, *The Origin of Humankind* (Basic Books 1996) pp. 102–103.

"Even the strongest modern professional wrestler": Ibid., (1996) pp. 102–103.

The extremely robust Boxgrove tibia appears to have belonged to such a person: Robin Dennell, "The world's oldest spears," *Nature 385.* Nature Publishing Group (1997) pp. 767 to 768.

an article by Denis Bramble and David Carrier in a 1983 issue of *Science*: Denis Bramble and David Carrier, "Running and breathing in mammals," *Science* 219 (4582) (Jan 21, 1983) pp. 251–256.

our venous circulation may be "designed" to carry blood cooled by sweating: Denis Bramble and Daniel Lieberman, "Endurance running and the evolution of *Homo*," *Nature* 432 (2004) pp. 345–352.

arterial blood on its way to the brain by way of the internal carotid artery: Dean Falk, "Evolution of cranial blood drainage in hominids: Enlarged occipital/marginal sinuses and emissary foramina," *American Journal of Physical Anthropology* Vol. 70 (July 1986) pp. 311–324.

had evolved in response to that species' strategy of pursuing swift, open-country prey: David R. Carrier, "The Energetic Paradox of Human Running and Hominid Evolution," *Current Anthropology* Vol. 25, No. 4 (New York: The Wenner-Gren Foundation for Anthropological Research 1984).

using endurance running to secure a wide variety of swift-running prey: Peter Nabokov, *Indian Running: Native American History and Tradition* (Ancient City Press June 1, 1987). See also Wendell C. Bennett and Robert M. Zingg, *The Tarahumara: An Indian Tribe of Northern Mexico* (Chicago: University of Chicago Press January 1, 1935).

until the animals were exhausted and then throttled them to death by hand: Bernd Heinrich, *Why We Run: A Natural History* (Manhattan: Ecco 2002) pp. 174–175.

Paiutes and Navajos were reputed to do the same with pronghorn antelopes: Op. cit., Heinrich, *Why We Run* (2001) p. 175.

was, in the early Pleistocene, still "tightly coupled": E. O. Wilson, *Consilience: The Unity of Knowledge* (New York: Vintage Books 1999) p. 182.

Oldowan assemblages undergo distinct refinement in chipping techniques: Roy Larick and Russell Ciochon, "The African Emergence and Early Asian Dispersals of the Genus Homo," *American Scientist* (AMER SCI. 84.) (1996) pp. 538–551.

Acheulean tools were dramatically bigger than their Oldowan counterparts, and more efficient: Nicholas Toth, "The Oldowan reassessed: A close look at early stone artifacts," *Journal of Archaeological Science,* Vol. 12, No. 2 (Amsterdam: Elsevier 1985) pp. 101–120.

reveal that those tools were, in fact, used as "knives" and "saws" to cut a variety of substances: Dominguez-Rodrigo, Serrallonga et al., "Woodworking activities by early humans: a plant residue analysis on Acheulian stone tools from Peninj (Tanzania)," *Journal of Human Evolution* Vol. 40, No. 4 (Amsterdam: Elsevier April 2001) pp. 289–299.

"one of the many enormous, unsightly, opencast brown-coal mines...": R. Dennell, "The world's oldest spears," *Nature* Vol. 385 (1997) p. 767.

and a great many flakes produced in the process of retouching those tools: Hartmut Thieme, "Die ältesten Speere der Welt— Fundplätze der frühen Altsteinzeit im Tagebau Schöningen," *Archäologisches Nachrichtenblatt* 10 (2005) pp. 409–417.

it was overshadowed by the discovery of sevel well-preserved wooden spears: Ibid., (2005) pp. 409–417.

"to regard them as snow-probes or digging-sticks is like claiming that power drills are paperweights": Op. cit., Dennell, *Nature* 385 (1997) p. 767.

Fully 60 percent of the enormously abundant bone frag-ments that hominins had processed and accumulated at Bilzingsleben: D. Mania, "Bilzingsleben—middle Pleisto-cene site of Homo erectus. Travertine complex and fauna at Bilzingsleben," *Quaternary field trips in Central Europe*, 14. Congress. INQUA Berlin (1995), pp. 738–740, pp. 777–780, pp. 1078–1079.

Each of those kills would, Roberts tells us, have been "a magnet for other predators": R. McKie, "Boxgrove Man Goes Back Underground," *The Observer* (October 20, 1996). See gener-ally Michael W. Pitts and Mark Roberts, *Fairweather Eden: life in Britain half a million years ago as revealed by the excavations at Boxgrove* (Salt Lake City: Century 1997).

I learned the art from a poor native who had nothing but a homemade spear: George Brommers, "Interviewing the Tiger-Man," *Ye Sylvan Archer* (May 1937).

He was also present in areas of Zimbabwe where thousands of elephants had been shot: Gary Haynes, "Taphonomic Studies of Elephant Mortality in Zimbabwe," *Elephant* 2(2). (Elephant Interest Group (EIG) 1986) pp. 69–71.

CHAPTER 13: FIRE

the discovery of convincing evidence of hominin fire-use: Paul Goldberg and Francesco Berna, "Microstratigraphic evidence of in situ fire in the Acheulean strata of Wonderwerk Cave, Northern Cape province, South Africa," *PNAS* No. 20 (2012) p. 109.

Members 1–3: The deposits are numbered in the order in which they appear to have accumulated in the cave, with Member 1 being the oldest, Member 2 being the second oldest, and so on.

I would have had no hesitation in assuming that it had been partly charred in a fire: C. K. Brain, "The Occurrence of Burned Bones at Swartkrans and Their Implications for the Control of Fire by Early Man, in Swartkrans," *A Cave's Chronicle of Early Man*, Second Edition, (Ed.) C. K. Brain. Transvaal Museum Monograph 8 (2004) p. 229.

In one grid square, W3/S3, burnt bones occur in 23 excavation spits: Op. cit., Brain (2004) p. 240.

Brain cut the fresh, defleshed radius of a hartebeest into seven transverse segments: Ibid., pp. 230–237.

subsequently confirmed these findings by using ESR to test samples of the bones: Anne R. Skinner, Joanna L. Lloyd, C. K. Brain, and F. Thackeray, *"Electron Spin Resonance and the First Use of Fire,"* Paleontology Society Meeting, Montreal, Quebec (March 2004).

approximate the Developed Oldowan of eastern Africa: Op. cit., Howell (2004) p. xi.

more likely to have been produced by digging in termite mounds: Lucinda Backwell and Francesco d'Errico, "Evidence of termite foraging by Swartkrans early hominids," *National Academy of Sciences* (Feb 13, 2001).

similar microscopic wear has been documented on experimentally made and used awls: Op. cit., (2004) pp. 202–203 and p. 213.

it was possible that both hominin species had made and used those tools: L. S. B. Leakey, P. V. Tobias and J. R. Napier. "A new species of the genus Homo from Olduvai Gorge," *Nature* 202 (1964) pp. 7–9.

Erectus is thought to have butchered giant gelada baboons later in the Pleistocene: Pat Shipman, Wendy Bosler, et al., "Butchering of Giant Geladas at an Acheulian Site," *Current Anthropology* Vol. 22 (1981) p. 257.

Paranthropus's hands were well-adapted to precision grasping and the use of tools: Randall Susman, "Hand function and tool behavior in early hominids," *Journal of Human Evolution* Vol. 35, No. 1 (July 1998) pp. 23–46.

the so-called "Hanging Remnant," was between 1.8 and 1.5 million years old: Elizabeth Vrba, "Biostratigraphy and chronology, based particularly on Bovidae, of the southern hominid-associated assemblages: Makapansgat, Sterkfontein, Taung, Kromdraai, Swartkrans; also Elandsfontein (Saldanha), Broken Hill (now Kabwe) and Cave of Hearths," (Eds.) H. de Lumley, and M.-A. de Lumley, *Pretirage, ler Cong. Internat. Paleo. Humaine* Vol. II (1982) pp. 707–752.

all three of the older Members were deposited in the neighborhood of 1.6 million years ago: D. J. de Ruiter, "Revised faunal lists for Members 1–3 of Swartkrans, South Africa," *Annals of the Transvaal Museum* 40(2003) p. 29 to p. 39.

an age of between 1.47 million years and 1.79 million years with a median of 1.63 million years: Darren Curnoe, Rainer Grün, et al., "Direct ESR dating of a Pliocene hominin from Swartkrans," *Journal of Human Evolution* Vol. 40 (2001) pp. 379–391.

a telling pattern of thermal alteration: Brian Ludwig, "New evidence for the possible use of controlled fire from ESA sites in the Olduvai and Turkana basins," *Abstracts for the Paleoanthropology Society Meetings* (Philadelphia: The University of Pennsylvania Museum 4–5 April 2000) p. A17.

a response that was initially skeptical but now tends toward acceptance: Sarah Hlubik, Russell Cutts, David Braun, Francesco Berna, Craig Feibel and John Harris, "Hominin fire use in the Okote member at Koobi Fora, Kenya: New evidence for the old debate," *Journal of Human Evolution* 133 (2019) p. 214.

tests of four of the discolored patches conducted in association with Charles Peters: Ralph Rowlett and Charles Peters, "Burnt earth associated with hominid site KxJj 20 East in Koobi Fora," Glynn Isaacs and Barbara Isaacs (Eds.) *Oxford University Press* Vol. 5 (1997).

several possible fireplaces at Koobi Fora's FxJi-20 main site: Randy Bellomo, "Methods of determining early hominid behavioral activities associated with the controlled use of fire at FxJj 20 Main, Koobi Fora, Kenya," *Journal of Human Evolution* Vol. 27 (1994) p. 173.

examined the phytoliths left by various kinds of fires: Ralph Rowlett, "Fire Control by *Homo erectus* in African and East Asia," *Acta Anthropologica Sinica*, supplement to Vol. 19 (2000) p. 198, pp. 201–204.

Hominin used fire may also have discolored the clumps of earth discovered in Chesowanja: Glynn Isaac, "Early hominids and fire at Chesowanja, Kenya," *Nature* Vol. 296 (1982) p. 870.

Burned seeds, wood, and flint found at Gesher Benot Ya'aqov: Nira Alperson-Afil, Daniel Richter and Naama Goren-Inbar, "Evaluating the intensity of fire at the Acheulian site of Gesher Benot Ya'aqov—Spatial and thermoluminescence analyses," *Science* Vol. 304 No. 5671 (30 April 2004) pp. 725–727.

when dry electric storms ignite grass or brush: W. Trollope and A. Potgieter, "Fire behaviour in the Kruger National Park," *Journal of the Grassland Society of Southern Africa* (1985).

CHAPTER 14: METAPHORICAL NEW ZEALAND

"[h]e has either decimated and [*sic*] eradicated the earth's animals": Op. cit., Raymond Dart, *Adventures* p. 198.

Most top predators of the planet are majestic creatures: Yuval Noah Harari, *Sapiens: A Brief History of Humankind,* (Canada: McClelland & Stewart 2014) p. 12.

"not even wrong": The physicist Wolfgang Pauli is credited with this response to incorrect or careless thinking. It is sometimes quoted as "That is not only not right; it is not even wrong," or in Pauli's native German, "Das ist nicht nur nicht richtig; es ist nicht einmal falsch!" https://en.wikipedia.org/wiki/Not_even_wrong (Retrieved April 5, 2021).

a group of hyenas killed, in one night, more than 110 Thomson's gazelle: Hans Kruuk, *The Spotted Hyena* (Chicago: University of Chicago Press 1972) p. 89.

younger lion were often in poor, even emaciated condition: James Stevenson-Hamilton, *South African Eden: the Kruger National Park, 1902–1946* (Cape Town, South Africa: Struik Publishers 1993) p. 74.

starvation and starvation-related killings: Fritz Eloff, *Hunters of the Dunes—The story of the Kalahari lion* (Pretoria, South Africa: Protea Publisher Nov. 29, 2016) 2nd Ed.

George Schaller, working in East Africa, came up with an estimate of one in five: George Schaller, *The Serengeti Lion: A Study of Predator-Prey Relations* (Chicago: University of Chicago Press 1976).

"Heaven and Earth are heartless": Lao Tzu, *Tao Te Ching* Verse Five: Living Virtuously: "Heaven and Earth are impartial; they treat all of creation as straw dogs." J. H. MacDonald (Translator) (London: Arcturus Editions 2018).

We behold the face of nature bright with gladness, we often see superabundance of food: Charles Darwin, *On the Origin of Species by Means of Natural Selection, or the Preservation of Favoured Races in the Struggle for Life* (London: John Murray 1859) 1st ed. p. 69.

Nothing is easier to admit in words than the truth of the universal struggle for life: Ibid., p. 62.

"largely unabsorbed into popular consciousness": Richard Dawkins, *Blind Watchmaker* (New York: Norton and Co. 1986) Preface, p. xix.

dominance over-developed can do damage of an absolute order: Robert Ardrey, *African Genesis. A Personal Investigation into the Animal Origins and Nature of Man* (London: Collins 1961) p. 105.

I well remember my conviction that there is more in man than the mere breath in his body: Charles Darwin, *The life and letters of Charles Darwin* (London: John Murray 1887) p. 311.

I was very unwilling to give up my belief: Darwin, *The life and letters*, pp. 308–309.

Some have attempted to explain this in reference to man: Darwin, *The life and letters*, p. 311.

we require at all times a certain quantity of care or sorrow or want, as a ship requires ballast, in order to keep on a straight course: Arthur Schopenhauer, *On the Suffering of the World* (Penguin Books 2004) p. 3.

"If you're not shocked by the theory of quantum mechanics":
Karen Barad, *Meeting the Universe Halfway: Quantum Physics and the Entanglement of Matter and Meaning* (Durham, NC: Duke University Press 2007) p. 254.

Scientists are first and foremost human beings, as subject to cultural bias as anyone: Bill McKibben and David Suzuki, *The David Suzuki Reader: A Lifetime of Ideas from a Leading Activist and Thinker* (Vancouver: Greystone Books/David Suzuki Foundation 2003) 1st ed. p. 202.

the "pacts" that have established a mutually dependent reproductive strategy: Richard Dawkins, *Climbing Mount Improbable* (London: Penguin 1997) p. 309.

the countless organisms in our biosphere that cooperate with each other: John Maynard Smith and Eörs Szathmáry, *The Origins of Life: From the Birth of Life to the Origin of Language* (Oxford: Oxford University Press 2000) p. 207.

Lynne Margulis, who marshaled the evidence that persuaded biologists: Lynne Margulis, *Acquiring Genomes: A Theory of the Origin of Species* (New York: Basic Books 2003) reprint ed.

plagues of such vastness that they have never since been rivaled: Op. cit., Flannery (2002) pp. 54–55.

CHAPTER 15: A RADICALLY DIFFERENT KIND OF FACULTY

as Richard Dawkins's 1997 *Climbing Mount Improbable* demonstrates so convincingly: Richard Dawkins, *Climbing Mount Improbable* (New York: W. W. Norton 1997).

"Why," William James asked, "does a particular maiden turn our wits so upside-down?: *Delphi Complete Works of William James* (Delphi Classics 2018) p. 387.

"What voluptuous thrill," James goes on to ask, "may not shake a fly, when she at last discovers the one particular leaf": James, *Complete Works,* pp. 387–388.

The first foundation or origin of the moral sense lies in the social sense: Charles Darwin, *The Descent of Man, and Selection in Relation to Sex* (London: John Murray 1871) p. 391.

Your father's opinion that all morality has grown up by evolution: Henrietta Litchfield, (Ed.). This letter appeared in *Emma Darwin, Wife of Charles Darwin. A Century of Family Letters* (London: John Murray/University Press 1904) pp. 360–361.

the breathtaking brilliance of the designs to be found in Nature: Daniel Dennett, *Darwin's Dangerous Idea: Evolution and the Meanings of Life* (New York: Simon and Schuster 1995) p. 74.

'Orgel's Second Rule: Evolution is cleverer than you are': Ibid. p. 74.

its phylogenetic "collaborator"—that is, natural selection—may have responded with significant changes of its own: Richard Wrangham, *Catching Fire: How Cooking Made Us Human* (New York: Basic Books 2010).

has invented and is able to use various weapons, tools, traps, etc.: Op. cit., Darwin (1871) p. 132.

the extinction of many prey species in whatever environments they have penetrated: J. Tooby and I. DeVore, "The Reconstruction of Hominid Behavioral Evolution Through Strategic Modeling," *The Evolution of Human Behavior: Primate Models,* (Ed.) Warren G. Kinzey (Albany: SUNY Press 1987) p. 183.

A female raven caged with others of her species in temporary captivity: Bernd Heinrich, *Mind of the Raven: Investigations and Adventures with Wolf-Birds* (New York: Harper Collins 2000), pp. 314 et seq.

New Caledonian crows modify and use twigs: G. R. Hunt and R. D. Gray, "Diversification and cumulative evolution in New Caledonian crow tool manufacture," *Proceedings of the Royal Society Biological Sciences* Vol. 270, No. 1517 (2003) pp. 867–874.

the sole occupants of Tooby and DeVore's "cognitive niche": Op. cit., Tooby and DeVore (1987) p. 183. See also Steven Pinker, "The cognitive niche: Coevolution of intelligence, sociality, and language," *PNAS 107* (Supplement 2) (May 11, 2010) pp. 8993–8999, and Andrew Whiten and David Erda, "The human socio-cognitive niche and its evolutionary origins," *Philosophical Transactions of the Royal Society Biological Sciences* Vol. 367 (2012) pp. 2119–2129.

***Darwin's Dangerous Idea...*"Vast" (which he defines as "very-much-more-than -usual"):** Op. cit., Dennett (1995) p. 109.

the readiness with which the imagination suggests its ideas and presents them at the very instant: David Hume, *Treatise of Human Nature,* in *The Essential David Hume* R. P. Wolf (Ed.) (New York: New American Library 1969) p. 48.

that our species has *more*, rather than less, instincts than other animals: William James, *The Principles of Psychology* (New York: Holt 1890) pp. 389 et seq.

proclamation of the Sabi Game Reserve: In 1846 the Volksraad (the parliament of the Transvaal Republic) proclaimed a law prohibiting hunting by foreigners and permitting game hunting by citizens only for own consumption. Eventually on 26 March 1898, the 'Sabiewildreserwe' was proclaimed. James Stevenson-Hamilton served from 1902 to 1946 as the first warden of South Africa's Sabi Nature Reserve, which was expanded to become Kruger National Park in 1926. Only in 1927 was the reserve opened to visitors.

[t]he large herds, containing females and young, are shy and keep away from the roads: Colonel Stevenson-Hamilton, *Our South African National Parks* (Cape Town, South Africa: Cape Times Limited 1940).

Do not stop to photograph a herd of elephants: Eric Robins, *Africa's wildlife: Survival or extinction?* (London: Odhams Press, 1 Jan 1963).

The white rhinoceros is a giant that weighs, on average, about twice as much as the black rhinoceros: It's widely repeated, these days, that the name "white rhino" is a "corruption" or "mistranslation" of the Dutch name *"wijde renoster."* The word *"wijde"* ("wide") is thought to refer to the white rhino's top lip. The grass-eating white rhino has a straight, lawn-mower-type top lip, while the tree- and shrub-eating black rhino's top lip is pointed into a kind of mini-trunk. This cannot, however, be right. Firstly, the Afrikaans name for the white rhinoceros isn't *"wijde renoster."* It is, in fact, *"wit renoster"* (which means, simply, "white rhinoceros"). And nobody has argued that the smaller species has ever been called anything else than *"swart renoster"* (black rhino).

to everyone's surprise...a handful of these giant rhinos had survived: Ian Player, *The white rhino saga* (London: Collins 1972).

There can, I fear, be little doubt that this rare and interesting quadruped: H. A. Bryden, *Nature and Sport in South Africa* (London: Chapman and Hall 1897) p. 190.

Guise wrote to the Governor of Natal...demanding that the animals be given protected status: 9 ZGH 762, Minute Paper Z13OM 895. "C. D. Guise Submits Suggestions in regard to the Preservation of Game in Zululand." C. D. Guise to Secretary for Zululand (19 February 1895).

population increases in both black and white rhino numbers in Kruger have given way to declines: S. M. Ferreira, C. Greaver, et al., "Disruption of Rhino Demography by Poachers May Lead to Population Declines in Kruger National Park, South Africa" Benjamin Lee Allen (Ed.) *PLoS ONE* Vol. 10, No. 6 (Brisbane, Australia: University of Queensland June 2015).

to make out the details of all the flies, bees, moths and butterflies that were trying: "...it seems probable that the number of insect species occurring in Kruger is somewhere between 40 and 60 percent of the total number of species present in South Africa (Braack, unpublished records). If the estimate is correct, then about 50% of South Africa's insect diversity is being conserved in less than 4% of the country." Leo Braak and Per Kryger, "Insects and Savanna Heterogeneity," *The Kruger Experience. Ecology and Management of Savanna Heterogeneity*, J. du Toit et al., (Eds.) (Washington, DC: Island Press 2003) (Chapter 12), p. 263.

Palgrave's *Trees of Southern Africa*: Keith Coates Palgrave, *Trees of Southern Africa* (Cape Town: Struik Publishers, 1977) 1st ed.

Codd's *Trees and Shrubs of the Kruger National Park*: L.E.W. Codd, *Trees and Shrubs of the Kruger National Park* (Pretoria, South Africa: Govt. Printer, 1951).

a book on fish and fishing in India: Henry Sullivan Thomas, *The Rod in India: Being Hints How to Obtain Sport, with Remarks on the Natural History of Fish, Their Culture, and Value; and Illustrations of Fish and Tackle* (London: Hamilton Adams and Co. 1898).

Rex Jubb's *Freshwater Fishes of Southern Africa*: R. A. Jubb, *Freshwater Fishes of Southern Africa* (Amsterdam and South Africa: A. A. Balkema 1967).

Bibliography

Alden, Peter, et al. *Collins guide to African Wildlife.* HarperCollins (2004).

Alexander, R. McNeill. *Principles of Animal Locomotion.* Princeton University Press (2002).

Alexeev, Valerii. *Smithsonian National Museum of Natural History: Human Evolution Evidence: Homo rudolfensis.* https://humanorigins.si.edu/evidence/human-fossils/species/homo-rudolfensis (retrieved April 6, 2021).

Allentoft, Morten Erik, Heller, Rasmus, et al. "Extinct New Zealand megafauna were not in decline before human colonization," *PNAS* Vol. 111 (13) (April 1, 2014) pp. 4922–4927.

Alperson-Afil, Nira, Richter, Daniel, Goren-Inbar, Naama. "Evaluating the intensity of fire at the Acheulian site of Gesher Benot Ya'aqov—Spatial and thermoluminescence analyses," *PLOS* (2017).

Alroy, J. "Putting North America's end-Pleistocene megafaunal extinction in context: large scale analyses of spatial patterns, extinction rates, and size distributions." *Extinctions in Near Time: Causes,*

Contexts, and Consequences. MacPhee, R., Sues, H., (Eds.) New York: Plenum Publishing (1999) pp. 105–143.

Anderson, Atholl, Clark, Geoffrey, et al. "New evidence of megafaunal bone damage indicates late colonization of Madagascar," *PLOS ONE*, Vol.13 (10) (2018).

Anton, Mauricio, Galobart, Àngel and Turner, Alan. "Co-existence of scimitar-toothed cats, lions and hominins in the European Pleistocene. Implications of the post-cranial anatomy of *Homotherium latidens* (Owen) for comparative palaeoecology," *Quaternary Science Reviews* 24 10–11; pp. 1287–1301 (2004).

Ardrey, Robert. *African Genesis. A Personal Investigation into the Animal Origins and Nature of Man.* Collins, London (1961).

Asfaw, Berhane, White, Tim, et al. "A New Species of Early Hominid from Ethiopia," *Science* Vol. 284, No. 5414 (1999).

Backwell, Lucinda and d'Errico, Francesco. "Evidence of termite foraging by Swartkrans early hominids," *National Academy of Sciences* (February 13, 2001).

Barad, Karen. *Meeting the Universe Halfway: Quantum Physics and the Entanglement of Matter and Meaning.* Duke University Press (2007).

Bather, F. A. "The Word 'Australopithecus' and Others," *Nature*, Vol. 115 (1925).

Bellomo, Randy. "Methods of determining early hominid behavioral activities associated with the controlled use of fire at FxJj 20 Main, Koobi Fora, Kenya," *Journal of Human Evolution* 27 (1994).

Bennett, Wendell C., and Zingg, Robert M. *The Tarahumara: An Indian Tribe of Northern Mexico.* The University of Chicago Press. (First Printing edition, January 1, 1935.)

Binford, Lewis. *Ancient Bones and Modern Myths.* Academic Press, Elsevier Inc. (1981).

Blondel, Jacques. "On humans and wildlife in Mediterranean islands," *Journal of Biogeography,* Vol. 35, Is. 3 (March 2008).

Boesch, Christophe and Hedwige. "Hunting behavior of wild chimpanzees in the Taï National Park," *American Journal of Physical Anthropology (78)* (1989).

Braak, Leo and Kryger, Per "Insects and Savanna Heterogeneity," *The Kruger Experience. Ecology and Management of Savanna Heterogeneity,* (Eds.) J. du Toit et al., Island Press (2003).

Brain, C. K. *Were our early ancestors murderers and headhunters? A prehistoric detective story,* Quest 5 (2). A joint project of the National Research Foundation and the Department of Science and Technology, Republic of South Africa (2009).

Brain, C. K. *The Hunters or the Hunted? An Introduction to African Cave Taphonomy.* University of Chicago Press (1981).

Brain, C. K. *A Century of Nature: Twenty-One Discoveries that Changed Science and the World* (Eds.) Laura Garwin and Tim Lincoln. University of Chicago Press, Chicago and London (2003).

Brain, C. K. *A Cave's Chronicle of Early Man,* Second Edition, (Ed.) C. K. Brain. Transvaal Museum Monograph 8 (2004).

Bramble, Denis and Carrier, David. "Running and breathing in mammals," *Science,* 219 (4582) (Jan 21, 1983).

Bramble, Denis and Lieberman, Daniel. "Endurance running and the evolution of Homo," Nature 432 (2004).

Braun, David R., Harris, John W. K., et al. "Early hominin diet included diverse terrestrial and aquatic animals 1.95 Ma in East Turkana, Kenya," (Ed) Richard G. Klein, PNAS 2010 107 (22). Stanford University, Stanford, CA (2010).

Brommers, George. "Interviewing the Tiger-Man," *Ye Sylvan Archer* (May 1937).

Brooks, Allison and McBrearty, Sally. "The Revolution that wasn't: a new interpretation of the origin of modern human behavior," *Journal of Human Evolution* (39) Academic Press (2000).

Brown, F, Harris, J, Leakey, R, Walker, A. "Early *Homo erectus* skeleton from west Lake Turkana, Kenya" *Nature* 316:6031 (1985).

Brunet, Michel, Beauvilain, A., et al. "The first australopithecine 2,500 kilometres west of the Rift Valley (Chad)," *Nature 16:378* (1995).

Bryden, H. A. *Nature and Sport in South Africa.* London Chapman and Hall (1897).

Bunn, Henry, et al. "FxJj50: An Early Pleistocene Site in Northern Kenya," *World Archaeology,* vol. 12, no. 2 (1980).

Carrier, David R. "The Energetic Paradox of Human Running and Hominid Evolution," *Current Anthropology* Vol. 25, No. 4. The Wenner-Gren Foundation for Anthropological Research (1984).

Cassella, Carly. "Plants Are Going Extinct at Least 500 Times Faster Than if Humans Weren't Around," https://www.sciencealert.com/ environment (12 June 2019).

Ceballos, Gerardo et al. "Accelerated modern human-induced species losses: Entering the sixth mass extinction," Science Advances Vol. 1, no. 5 (19 Jun 2015).

Chomsky, N. *Powers and Prospects. Reflections on human nature and the social order* (London, Pluto Press, 1996).

Clarkson, Chris, Jacobs, Zenobia, et al. "Human occupation of northern Australia by 65,000 years ago," *Nature* 547, 76630 (20 July 2017).

Codd, L. E. W. *Trees and Shrubs of the Kruger National Park.* Pretoria: Government Printer (1951).

Cremo, Michael A. *Searching for the truth with...The Forbidden Archaeologist: The Atlantis Rising Magazine Columns of Michael Cremo.* Los Angeles, Bhaktivedanta Book Publishing, Inc. (2010).

Crerar, Andrew P., et al. "Rewriting the history of an extinction—was a population of Steller's sea cows (*Hydrodamalis gigas*) at St Lawrence Island also driven to extinction?" *The Royal Society—Palaeontology* (2014).

Curnoe, Darren, Grün, Rainer et al. "Direct ESR dating of a Pliocene hominin from Swartkrans," *Journal of Human Evolution*, Vol. 40 (2001).

Currier, Richard. *Unbound: How Eight Technologies Made Us Human, Transformed Society, and Brought Our World to the Brink.* Arcade, Reprint edition (2015).

Dampier, William. *A new voyage round the world.* Vol. 1 [of Dampier's voyages.]. James Knapton, Australia (1703).

Dart, R. "Australopithecus africanus The Man-Ape of South Africa." *Nature* 115 (1925).

Dart, R. "The Taungs Skull," *Nature 116, 462* (1925).

Dart, R. "The predatory transition from ape to Man," *International Anthropological and Linguistic Review,* Vol. 1, No. 4. Brill (1953).

Dart, R. *Adventures with the missing link.* Hamish Hamilton (1959).

Darwin, Charles. *The Beagle Diary* (original 1834) Cambridge, Cambridge University Press (First edition 1988).

Darwin, Charles. *Journal of Researches Into the Geology and Natural History of the Various Countries Visited by HMS Beagle* (1840).

Darwin, Charles. *"On the Origin of Species by Means of Natural Selection, or the Preservation of Favoured Races in the Struggle for Life"* (London: John Murray) (First edition, 1859).

Darwin, Charles. *"The Descent of Man, and Selection in Relation to Sex."* London, John Murray (1871).

Darwin, Charles. *Journal of researches into the natural history and geology of the various countries visited by H.M.S. Beagle etc.* London, John Murray (1890).

Darwin, Charles. *The Works of Charles Darwin, Vol. 12: A Monograph of the Sub-Class Cirripedia, Vol. II: The Balanidae (Part One)* NYU Press (1989).

Darwin, Francis. *The life and letters of Charles Darwin.* London, John Murray (1887).

Dawkins, Richard. *Blind Watchmaker.* Norton and Co. (1986).

Dawkins, Richard. *Climbing Mount Improbable.* W. W. Norton and Company (1997).

Dawkins, Richard. *A Devil's Chaplain: Reflections on Hope, Lies, Science, and Love.* Boston: Houghton Mifflin (2013).

Dayal, Manisha, et al. "The History and Composition of the Raymond A. Dart Collection of Human Skeletons at the University of the Witwatersrand, Johannesburg, South Africa." *American Journal of Physical Anthropology* Vol. 140 (2009).

De Groote, Flink, et al. *New genetic and morphological evidence suggests a single hoaxer created 'Piltdown man.'* Royal Society Open Science (2016).

De Heinzelin, Jean, Clark, J. Desmond, White, Tim et al. "Environment and Behavior of 2.5-Million-Year-Old Bouri Hominids." *Science* Vol. 284, No. 5414 (1999).

De Ruiter, D. J. "Revised faunal lists for Members 1–3 of Swartkrans, South Africa." *Annals of the Transvaal Museum* 40 (2003).

Dennell, Robin. "The world's oldest spears," *Nature* 385 (1997).

Dennell, Robin. *The Palaeolithic Settlement of Asia.* Cambridge World Archaeology, Cambridge: Cambridge University Press (2008).

Dennett, Daniel. *"Darwin's Dangerous Idea: Evolution and the Meanings of Life."* Simon and Schuster (1995).

Diamond, Jared. *Germs, Guns, and Steel: The fates of human societies.* W. W. Norton (1997).

Diamond, Jared. *The Third Chimpanzee: The Evolution and Future of the Human Animal* (New York, Harper Collins 1992) p. 328.

Dominguez-Rodrigo, M., Serrallonga, J. et al. "Woodworking activities by early humans: a plant residue analysis on Acheulian stone tools from Peninj (Tanzania)," *Journal of Human Evolution* Vol. 40, No. 4. Elsevier April (2001).

Dominguez-Rodrigo, Manuel, Pickering, Travis, et al. "Cut-marked bones from Pliocene archaeological sites at Gona, Afar, Ethiopia: Implications for the function of the world's oldest stone tools," *Journal of Human Evolution* Vol. 48, No. 2 (2005).

Domning, Daryl P., et al. "Steller's Sea Cow in the Aleutian Islands," *Marine Mammal Science.* Society for Marine Mammalogy (2007).

Dorst, Jean, and Dandelot, Pierre. *A Field Guide to the Larger Mammals of Africa Hardcover.* Collins; 2nd edition (1978).

Eastham, Anne and Michael. "Palaeolithic images and the Great Auk," Antiquity; Gloucester Vol. 69, No. 266 (1995).

Eloff, Fritz. *Hunters of the Dunes: The Story of the Kalahari Lion.* Sunbird Publishers (1st edition 2002).

Estes, Richard D. *The Safari Companion: A Guide to Watching African Mammals.* Chelsea Green Publishing (1999).

Falk, Dean. "Evolution of cranial blood drainage in hominids: Enlarged occipital/marginal sinuses and emissary foramina." *American Journal of Physical Anthropology,* (1986).

Falk, Dean. "Brain evolution in Homo: The 'radiator' theory," *Behavioural and Brain Sciences* 13 (1990).

Falk, Dean. *Fossil Chronicles—How Two Controversial Discoveries Changed Our View Of Human Evolution.* University of California Press (2011).

Ferreira, S. M., Greaver, C. G., Knight, A., Knight, M. H., Smit, I. P. J., Pienaar, Danie. "Disruption of Rhino Demography by Poachers May Lead to Population Declines in Kruger National Park, South Africa," (Ed.) Benjamin Lee Allen *PLoS ONE* 10(6) (June 2015).

Flannery, Tim. *The Future Eaters: An Ecological History of the Australasian Lands and People.* Grove Press (2002).

Fowler, Brenda. "Scientists at Work: Schick, Kathy and Toth, Nicholas. "Recreating Stone Tools to Learn Makers' Ways," *The New York Times* (Dec. 20, 1994).

Gammage, Bill. *The Biggest Estate on Earth* (Crows Nest, Australia: Allen and Unwin, 2011).

Garwin, Laura, and Lincoln, Tim (Eds.). *A Century of Nature: Twenty-One Discoveries that Changed Science and the World* (Chicago and London: University of Chicago Press, 2003).

Gibbons, Ann. "Spear-Wielding Chimps Seen Hunting Bush Babies," *Science* 315 (2007).

Goldberg, Paul, and Berna, Francesco. "Microstratigraphic evidence of in situ fire in the Acheulean strata of Wonderwerk Cave, Northern Cape province, South Africa," *PNAS* (20) (2012).

Goodall, Jane. "Jane Goodall Reporting from Gombe Stream Game Reserve, 8 Oct 1962: 'Dear Mr. Payne...Twice again during the month I have seen them with meat...once the prey was almost certainly a red colobus monkey...'" *National Geographic Resource Library: Article* (1962).

Gould, Stephen Jay. *"Hen's Teeth and Horse's Toes: Further Reflections in Natural History."* W. W. Norton (1994).

Grün, R., Brink, J., Spooner, N. et al. "Direct dating of Florisbad hominid," *Nature* 382 (1996).

Guise, C. D. to Secretary for Zululand, "C. D. Guise Submits Suggestions in regard to the Preservation of Game in Zululand," 9 ZGH 762, Minute Paper Z13OM 895 (19 February 1895).

Häckel, Ernst. *Natürliche Schöpfungsgeschichte* (monograph) Berlin: Reimer (1868).

Haltenorth, Theodore, and Diller, Helmut. *Mammals of Africa including Madagascar,* English translation. Collins (1984).

Harari, Yuval Noah. *Sapiens: A Brief History of Humankind, Homo Deus.* Random House Canada (2015).

Harmand, Sonia, Lewis, Jason, Feibel, Craig, et al. "3.3-million-year-old stone tools from Lomekwi 3, West Turkana, Kenya," *Nature* 521 (2015).

Harrison, Terry. *Paleontology and Geology of Laetoli: Human Evolution in Context: Vol 2: Fossil Hominins and the Associated Fauna,* Springer Science and Business Media (2011).

Hart, Donna, Sussman, Robert Wald, et al. *Man the Hunted: Primates, Predators, and Human Evolution* (Boulder: Westview Press 2005).

Harvati, Katerina. "100 years of Homo heidelbergensis—life and times of a controversial taxon," *Mitteilungen der Gesellschaft für Urgeschichte 16* (2007).

Haynes, Gary. "Taphonomic Studies of Elephant Mortality in Zimbabwe," *Elephant,* 2(2) (Elephant Interest Group (EIG) 1986.

Heinrich, Bernd. *Mind of the Raven: Investigations and Adventures with Wolf-Birds,* HarperCollins (2000).

Heinrich, Bernd. *Why We Run: A Natural History,* Ecco (2002).

Herridge, Victoria L., and Lister, Adrian M. "Extreme insular dwarfism evolved in a mammoth," *Proc. R. Soc. B.*279 (2012).

Heymann, Eckhard W., et al. *South American Primates: Comparative Perspectives in the Study of Behavior, Ecology, and Conservation.* Springer Science and Business Media (2008).

Hlubik, Sarah, Cutts, Russell, et al. "Hominin fire use in the Okote member at Koobi Fora, Kenya: New evidence for the old debate," *Journal of Human Evolution,* 133 (2019).

Holloway, Ralph L. "The Evolution of the Hominid Brain," *Handbook of Paleoanthropology.* Springer, Berlin, Heidelberg (2015).

Howell, F. C. Foreword to *A Cave's Chronicle of Early Man, A Cave's Chronicle of Early Man,* Second Edition, (Ed.) C. K. Brain. Transvaal Museum Monograph 8 (2004).

Hume, David. *Treatise of Human Nature,* in *The Essential David Hume,* (Ed.) R. P. Wolf. New American Library (1969).

Hunt, G. R., and Gray, R. D. "Diversification and cumulative evolution in New Caledonian crow tool manufacture." *Proceedings of the Royal Society Biological Sciences* Vol. 270, No. 1517 (2003).

Isaac, Glynn. "Early hominids and fire at Chesowanja, Kenya," *Nature* (1982) 296.

James, William. *The Principles of Psychology.* Holt: New York (1890).

James, William. *Delphi Complete Works of William James.* Delphi Classics (2018).

Janzen, Daniel H., and Martin, Paul S. "Neotropical Anachronisms: The Fruits the Gomphotheres Ate." *Science,* vol. 215 no.4528 (1982) pp. 19–27.

Jensen, Mari N. "Paul S. Martin, Pleistocene Extinctions Expert, Dies," *College of Science, University of Arizona News* (Sept. 16, 2010).

Johanson, Donald, and Edgar, Blake. *From Lucy to Language*, New York: Simon & Schuster (1996).

Johanson, Donald C., and Edey, Maitland Armstrong. *Lucy, the Beginnings of Humankind,* Simon and Schuster (1981).

Jubb, R. A. *Freshwater Fishes of Southern Africa,* A.A. Balkema, Amsterdam and South Africa (1967).

Jukarab, A. M., Lyons, S. K. et al. "Palaeogeography, Palaeoclimatology, Palaeoecology" *Late Quaternary extinctions in the Indian Subcontinent,* Vol. 562 (Jan. 15, 2021).

Keast, Allen, *Evolution, Mammals, and Southern Continents* Chicago, IL: University of Chicago Press Journals, 1972).

Kathrein, Jonathan, and Kathrein, Margaret. *Surviving the Shark: How a Brutal Great White Attack Turned a Surfer into a Dedicated Defender of Sharks.* Skyhorse Publishing (2012).

Keeley, L., and Toth, N. "Microwear polishes on early stone tools from Koobi Fora, Kenya," *Nature* 293 (1981).

Keith, Sir Arthur. *The antiquity of man,* Williams and Norgate (1915).

Keith, Sir Arthur. "The Fossil Anthropoid Ape from Taungs," *Nature* Vol. 115 (1925).

Keith, Sir Arthur. "The Taungs Skull" *Nature* Vol. 116 (1925).

Keith, Sir Arthur. *New Discoveries Relating to the Antiquity of Man.* Williams and Norgate (1931).

Keith, Sir Arthur. *An Autobiography,* Philosophical Library, First Edition (1950).

Klein, Richard G., and Cruz-Uribe, Kathryn. "Middle and Later Stone Age large mammal and tortoise remains from Die Kelders Cave 1, Western Cape Province, South Africa," *Journal of Human Evolution* (2000).

Kovarovic, Kristin. *Bovids as Palaeoenvironmental Indicators*, Thesis. University College London (2004).

Kruuk, Hans. *The Spotted Hyena*. University of Chicago Press, Chicago (1972).

Kurtén, Björn. *The Ice Age*. Hart-Davis (1972).

Larick, Roy, and Ciochon, Russell. "The African Emergence and Early Asian Dispersals of the Genus Homo," *American Scientist* (AMER SCI. 84.) (1996).

Leakey, L. S. B., Tobias, P. V., et al. "A new species of the genus Homo from Olduvai Gorge, Tanzania," *Nature* 202 (1964).

Leakey, Meave, Feibel, C. S. et al. "New specimens and confirmation of an early age for *Australopithecus anamensis*," *Nature* 393 (1998).

Leakey, Meave, Spoor, F. et al. "New hominin genus from eastern Africa shows diverse middle Pliocene lineages," *Nature* 410 (2001).

Leakey, R. E. F., and Walker, Alan C. "Australopithecus, Homo erectus and the single species hypothesis," *Nature* Vol. 261 (1976).

Leakey, Richard, and Lewin, Roger. *Patterns of Life and the Future of Humankind*. Doubleday (1995).

Leakey, Richard. *The Origins of Humankind* (Basic Books 2008).

Lewis, M. E., and Werdelin, L. "Patterns of change in the Plio-Pleisto-cene carnivorans of eastern Africa," in *Hominin Environments in the East African Pliocene: An Assessment of the Faunal Evidence,* (Eds.) R. Bobe et al. (Springer, Dordrecht 2007).

Litchfield, Henrietta (Ed.). *Emma Darwin, Wife of Charles Darwin. A Century of Family Letters,* University Press (1904).

Louys, Julien, Curnoe, Darren et al. "Characteristics of Pleistocene megafauna extinctions in Southeast Asia," *PALAEO—Palaeogeog-raphy, Palaeoclimatology, Palaeoecology* 243 (2007).

Ludwig, Brian. "New evidence for the possible use of controlled fire from ESA sites in the Olduvai and Turkana basins," *Abstracts for the Paleoanthropology Society Meetings.* The University of Pennsyl-vania Museum, Philadelphia, Pennsylvania, USA (4–5 April 2000).

Lydeard, Charles, et al. "The Global Decline of Nonmarine Mollusks," *BioScience* Vol. 54, No. 4 (April 2004) pp. 321–330.

Lyell, Charles. *The Principles of Geology: Being an Attempt to Explain the Former Changes of the Earth's Surface, by Reference to Causes now in Operation.* Three vols. (London, John Murray 1833).

MacPhee, Ross D. E., and Sues, Hans-Dieter. *Extinctions in Near Time: Causes, Contexts, and Consequences* (Boston: Springer 1999).

Makin v. Attorney General of New South Wales: [1894] AC 57, [1893] UKPC 56.

Mania, D. "Bilzingsleben—middle Pleistocene site of Homo erectus. Travertine complex and fauna at Bilzingsleben," *Quaternary field trips in Central Europe,* 14. Congress. INQUA Berlin (1995).

Mania, Dietrich, Fischer, Karlheinz, Weber, Thomas. *Bilzingsleben: Homo erectus, seine Kultur und seine Umwelt,* Volume 4. Deutscher Verlag der Wissenschaften, Bilzingsleben, East Germany (1983).

Marchant, James. "Alfred Russel Wallace: Letters and Reminiscences, Vol. 1 (of 2), *The Project Gutenberg EBook*. Digital and Multimedia Center, Michigan State University Libraries (2005).

Margulis, Lynne. *Acquiring Genomes: A Theory of the Origin of Species* (New York: Basic Books 2003) reprint ed.

Marks, Jonathan. "Why were the first anthropologists creationists?" *Evolutionary Anthropology* (11 January 2011).

Martin, P. S. "Africa and Pleistocene Overkill," *Nature 212* (1966).

Martin, P. S. and Wright, H. E. (Eds.). "Prehistoric overkill" in *Pleistocene Extinctions: The Search for a Cause* (New Haven, CT: Yale University Press, 1967).

Martin, P. S., Long, A., and Hansen R. Met al.. "Extinction of the Shasta ground sloth," *Geological Society of America Bulletin* Vol. 85 (1974).

Martin, P.S., and Long, A. "Death of American Ground Sloths," *Science,* Vol. 186, (1974).

Martin, P. S., and Mosimann, J. E. "Simulating overkill by paleoindians: did man hunt the giant mammals of the New World to extinction? Mathematical models show that the hypothesis is feasible," *American Scientist,* Vol. 63 (1975).

Martin, P. S., Hopkins, D. M., et al. (Eds.). "The pattern and meaning of Holarctic mammoth extinctions," *Paleoecology of Beringia* (New York: Academic Press, 1984).

Martin, P. S., and Klein, R. G. (Eds.). "Prehistoric overkill: the global model," *Quaternary Extinctions: A Prehistoric Revolution* (Tucson, Arizona: University of Arizona Press, 1984).

Martin, P. S., Steadman, D. W., et al. "Asynchronous extinction of late Quaternary sloths on continents and islands," *Proceedings of the National Academy of Sciences USA 102* (1999).

Martin, P. S., and Burney D. "Wild Earth," Reprinted in *Wild Earth: Wild Ideas for a World Out of Balance* (T. Butler, ed.) (Minneapolis, MN: Milkweed Editions, 2002).

Martin, P. S., and Steadman, D. W. "Extinction of birds in the late Pleistocene of North America," *Quaternary Extinctions: A Prehistoric Revolution* (P. S. Martin and R. G. Klein, Eds.) (Tucson: University of Arizona Press, 2005).

Martin, P. S. *Twilight of the Mammoths: Ice Age Extinctions and the Rewilding of America* (Berkeley: University of California Press, 2007).

Martinez, Gustavo. "'Fish-Tail' projectile points and megamammals: New evidence from Paso Otero 5 (Argentina)," *Antiquity* 75 (289) (2001).

Mazak. *Vratislav Zeitschrift fur Säugertierkunde* 35 (1970).

McDonald, H. Gregory. "More Than Just Horses—Hagerman Fossil Beds", *Rocks & Minerals*, 68:5 (1993).

McKibben, Bill, and Suzuki, David. *The David Suzuki Reader: A Lifetime of Ideas from a Leading Activist and Thinker*. Greystone Books/ David Suzuki Foundation (1st edition, 2003).

McKie, R. "Boxgrove Man Goes Back Underground," *The Observer* (October 20, 1996).

McNeill, Alexander R., *Principles of Animal Locomotion* (Princeton, NJ: Princeton University Press, 2002).

McPherron, S., Alemseged, Z., Marean, C. et al. "Evidence for stone-tool-assisted consumption of animal tissues before 3.39 million years ago at Dikika, Ethiopia" *Nature 466* (2010).

Menez, Alex. "The Gibraltar Skull: early history, 1848–1868," *Archives of natural history 45.1*. Edinburgh University Press, © The Society for the History of Natural History (2018).

Meredith, Martin. *Born in Africa: The Quest for the Origins of Human Life*. Public Affairs, the Perseus Group (2011).

Miller, Gifford H., and Brigham-Grette, Julie. "Amino acid geochronology: Resolution and precision in carbonate fossils," *Quaternary International, Volume 1*. Elsevier Ltd. (1989).

Morwood, Mike, and Oosterzee, Penny. *A New Human, the Startling Discovery and Strange Story of the "Hobbits" of Flores, Indonesia*. Harper Collins (2007).

Nabokov, Peter. *Indian Running: Native American History and Tradition*. (Ancient City Press, June 1, 1987).

Napolitano, Matthew, DiNapoli, F., Robert J., et al. "Reevaluating human colonization of the Caribbean using chronometric hygiene and Bayesian modeling," *Science Advances* Vol. 5, No. 12 (18 Dec 2019).

Nelson, Richard. "Searching for the Lost Arrow: Physical and Spiritual Ecology in the Hunter's World," (Eds.) Stephen H. Kellert, Edward O. Wilson. *The Biophilia Hypothesis* Island Press, Washington, DC (1993).

Olson, Storrs L., and James, Helen F. "The role of Polynesians in the extinction of the avifauna of the Hawaiian Islands," *Quaternary extinctions: A prehistoric revolution*, (Eds.) P. S. Martin and R. G. Klein. Tucson, University of Arizona Press (1984).

Paddayya, K. "The Epistemology of Archaeology: A Postscript to the New Archaeology." *Bulletin of the Deccan College Research Institute*, vol. 45 (1986).

Palgrave, Keith Coates. *Trees of Southern Africa*, Struik Publishers Cape Town (First published 1977).

Pickford, Martin and Senut, Brigitte. "The geological and faunal context of Late Miocene hominid remains from Lukeino, Kenya," *Comptes Rendus de l'Académie des Sciences—Series IIA—Earth and Planetary Science. 332* (2001).

Pika, Simone, et al. "Wild chimpanzees (*Pan troglodytes troglodytes*) exploit tortoises (*Kinixys erosa*) via percussive technology," *Scientific Reports* 9, Article number: 7661 (2019).

Pinker, Steven. *The Language Instinct,* (New York: Harper Perennial Modern Classics (2007).

Pinker, Steven. "The cognitive niche: Coevolution of intelligence, sociality, and language," *PNAS 107 (Supplement 2)* (May 11, 2010).

Pitts, Michael, and Roberts, W. Mark. *Fairweather Eden: life in Britain half a million years ago as revealed by the excavations at Boxgrove.* Century (1st Edition, 1997).

Player, Ian. *The White Rhino Saga,* London, Collins (1972).

Plotz, Roan David. *The interspecific relationships of black rhinoceros (Diceros bicornis) in Hluhluwe-iMfolozi Park*, a thesis submitted to Victoria University of Wellington in fulfilment of the requirement for the degree of Doctor of Philosophy in Ecology and Biodiversity.

Plummer, Thomas. "Flaked Stones and Old Bones: Biological and Cultural Evolution at the Dawn of Technology," *American Journal of Physical Anthropology* (2004).

Potts, R., and Shipman, P. "Cutmarks made by stone tools on bones from Olduvai Gorge, Tanzania." *Nature* 291 (1981).

Rizal, Y., Westaway, K. E., Zaim, Y. et al. "Last appearance of *Homo erectus* at Ngandong, Java, 117,000–108,000 years ago," *Nature 577* (2020).

Roberts, R. G., Flannery, Timothy F., et al. "New Age for the Last Australian Megafauna: Continent-wide extinction about 46,000 years ago," *Science* Vol. 292 (2001).

Robins, Eric. *Africa's wildlife: Survival or extinction?* Odhams Press, London (1 Jan 1963).

Rowlett, Ralph, and Peters, Charles. "Burnt earth associated with hominid site KxJj 20 East in Koobi Fora," (Eds.) Glynn Isaacs and Barbara Isaacs. *Oxford University Press, vol. 5* (1997).

Rowlett, Ralph. "Fire Control by *Homo erectus* in African and East Asia," *Acta Anthropologica Sinica*, supplement to vol. 19, (2000).

Ruspoli, Mario. *Cave of Lascaux.* Harry N. Abrams (Reissue edition 1987).

Schaller, George B. *The Serengeti Lion: A Study of Predator-Prey Relations (Wildlife Behavior and Ecology series)* (Chicago:, University of Chicago Press, March 15, 1976) (reprint ed.).

Schöningen. Roman-Germanic Central Museum Mainz (2012).

Schopenhauer, Arthur. *On the Suffering of the World.* Penguin Books (2004).

Schüle, Wilhelm. "Human evolution, animal behavior, and quaternary extinctions: Schüle, A paleo-ecology of hunting," *Homo* 41 (3) (1990).

Schüle, Wilhelm. "Landscapes and climate in prehistory: interaction of wildlife, man and fire," J.G. Goldammer (Ed.) *Fire in tropical biota, ecosystem processes and global challenges*. Berlin, Heidelberg, New York: Springer Verlag (1990).

Schüle, Wilhelm. "Mammals, Vegetation and the Initial Human Settlement of the Mediterranean Islands: A Palaeoecological Approach," *Journal of Biogeography,* Vol. 20 (1993).

Schüle, Wilhelm. "Prähistorischer Faunenschwund: Ursache und Wirkung des Absterbens," *Verlag Wissenschaft und Öffentlichkeit,* Freiburg im Breisgau (2001).

Schüle, Wilhelm, und Schuster, Sabine. "Klima, Speer und Feuer. Zur ökologischen Rolle das Frühen Menschen." *Jahrbuch des Römisch-Germanischen Zentralmuseums," Mainz 42* (1996).

Schweiger, Andreas H., and Svenning, Jens—Christian. "Down—sizing of dung beetle assemblages over the last 53,000 years is consistent with a dominant effect of megafauna losses," *OIKOS Journal: Synthesizing Ecology.* Nordic Society Oikos (2018).

Shattock, S. G. "Morbid thickening of the calvaria, and the reconstruction of bone once abnormal: A pathological basis for the study of the thickening observed in certain Pleistocene crania." *XXVIIth International congress of Medicine Section VII.* Published in London (July 1913).

Shipman, Pat. *The Man Who Found the Missing Link: Eugene Dubois and His Lifelong Quest to Prove Darwin Right*, Harvard University Press (2002).

Shipman, Pat, et al. "Butchering of Giant Geladas at an Acheulian Site," *Current Anthropology* Vol. 22 (1981).

Shreeve, Jamie. "'Little Foot' Fossil Skeleton Rivals Famous Lucy in Age," *National Geographic* (April 1, 2015).

Skinner, Anne R., et al. *"Electron Spin Resonance and the First Use of Fire,"* Paleontology Society Meeting, Montreal, Quebec (March 2004).

Smith, Grafton Elliot. *"Essays on The Evolution of Man."* London: Oxford University Press (1924).

Smith, John Maynard, and Szathmáry, Eörs. *The Origins of Life: From the Birth of Life to the Origin of Language.* OUP Oxford (2000).

Stanford, Craig B. "Wild Chimpanzees: Implications for the Evolutionary Ecology of Pliocene Hominids," *American Anthropologist* New Series, Vol. 98, No. 1, Wiley on behalf of the American Anthropological Association (1996).

Stanford, Craig B. *Chimpanzee and Red Colobus: The Ecology of Predator and Prey,* Harvard University Press (1998).

Steele, Teresa E. "A unique hominin menu dated to 1.95 million years ago," *PNAS* June 15, 2010, 107 (24).

Steller, Georg Wilhelm. *De Bestiis Marinis.* St. Petersburg Imperial Academy of Sciences (1751).

Stevenson-Hamilton, James. *Our South African National Parks,* Cape Times Limited, Cape Town, South Africa (1940).

Stevenson-Hamilton, James. *South African Eden: the Kruger National Park, 1902–1946,* Struik Publishers (1993).

Stiner, M. C., et al. "The tortoise and the hare. Small-game use, the broad-spectrum revolution, and paleolithic demography," *Curr Anthropol,* 41.

Stringer, Chris. *Lone Survivors: How We Came to Be the Only Humans on Earth* (New York: Macmillan, 2012).

Surbeck, Martin, and Hohmann, Gottfried. "Primate hunting by bonobos at LuiKotale, Salonga National Park." *ScienceDirect,* Volume 18 (2008).

Surovell, Todd A., et al. "Global evidence for proboscidean overkill," *PNAS* (April 26, 2005).

Surovell, Todd A., et al. "Test of Martin's overkill hypothesis using radio-carbon dates on extinct megafauna," *PNAS* Vol. 113 (4) (January 26, 2016).

Susman, Randall. "Hand function and tool behavior in early hominids," *Journal of Human Evolution* Vol. 35, No. 1 (July 1998).

Suzuki, David. *The Sacred Balance: Rediscovering Our Place in Nature* (Vancouver: Greystone Books, 2007).

Swisher, Carl C. III, et al. *Java Man: How Two Geologists Changed Our Understanding of Human Evolution,* University of Chicago Press (2001).

Tanner, Nancy, and Zihlman, Adrienne. "Women in Evolution," *Part I: Innovation and Selection in Human Origins* Vol. 1, No. 3., The University of Chicago Press (1976).

Tattersall, Ian. *The Fossil Trail: How We Know what We Think We Know about Human Evolution,* Oxford University Press (1995).

Tattersall, Ian. *Becoming Human: Evolution and Human Uniqueness,* Houghton Mifflin Harcourt (1999).

Teaford, Mark F., and Ungar, Peter S. "Diet and the evolution of the earliest human ancestors," *PNAS* Vol. 97 (25) (2000).

Theunissen, L. T. *Eugène Dubois and the Ape-Man from Java: The History of the First 'Missing Link' and Its Discoverer,* Springer Science and Business Media B.V. (1988).

Thieme, Hartmut. "Die ältesten Speere der Welt—Fundplätze der frühen Altsteinzeit im Tagebau Schöningen," *Archäologisches Nachrichtenblatt* 10 (2005).

Thieme, Hartmut, and Mania, D. "Schöningen (Northern Harz foothills). An ancient Paleolithic Find from the Middle Ice Age" Karl-Ernst Behre (Ed.) *Forshungen zur Urgeschichte aus dem Tagebau von Schöningen,* Publisher of the Roman-Germanic Central Museum Mainz, 2012.

Thomas, Henry Sullivan. *The Rod in India: Being Hints How to Obtain Sport, with Remarks on the Natural History of Fish, Their Culture, and Value; and Illustrations of Fish and Tackle,* London: Hamilton Adams and Co. (1898).

Tilde, Chris and Stuart. *Chris and Stuart Tilde's Field Guide to the Mammals of Southern Africa,* Ralph Curtis Publishing (1989).

Tobias, Phillip V. *Yearbook of Physical Anthropology, Vol. 28,* History of Physical Anthropology in Southern Africa (1985).

Tobias, Phillip V. "An Appraisal of the Case Against Sir Arthur Keith," *Current Anthropology ,* (1992).

Todd, N. E. "Trends in proboscidean diversity in the African Cenozoic," *Journal of Mammalian Evolution,* 13 (2006).

Tooby, J., and Cosmides, L. "Unravelling the Enigma of Human Intelligence. Evolutionary Psychology and the Multimodular Mind," in *Evolutionary Intelligence,* (Eds.) Robert J. Sternberg, James C. Kaufman, Psychology Press, NY (2013).

Tooby, J., and DeVore, I. "The Reconstruction of Hominid Behavioral Evolution Through Strategic Modeling," *The Evolution of Human Behavior: Primate Models,* (Ed.) Warren G. Kinzey, SUNY Press, Albany, NY (1987).

Toth, Nicholas. "Archaeological evidence for preferential right-handed-ness in the lower and middle Pleistocene, and its possible implica-tions," *Journal of Human Evolution* Vol. 14, No. 6, Elsevier (1985).

Toth, Nicholas. "The Oldowan reassessed: A close look at early stone artifacts," *Journal of Archaeological Science*, Vol. 12, No. 2 (Else-vier)(1985).

Trollope, W., and Potgieter, A. "Fire behaviour in the Kruger National Park," *Journal of the Grassland Society of Southern Africa* (1985).

Tsu, Lao. *Tao Te Ching* Verse Five: Living Virtuously: "Heaven and Earth are impartial; they treat all of creation as straw dogs." J. H. MacDonald (Translator) (London: Arcturus Editions, 2018).

Tudge, Colin. *The Time Before History: 5 Million Years of Human Impact,* Scribner Book Company (January 1, 1996).

Turner, Alan, and Anton, Mauricio. *The giant hyaena, Pachycrocuta brevirostris (Mammalia, Carnivora, Hyaenidae)* Geobios 29 (1996).

Turner, Alan, et al. "Changing ideas about the evolution and functional morphology of Machairodontine felids," *Estudios Geológicos* (2011).

Turney, C. S. M. "Development of a Robust 14C chronology for Lynch's Crater (North Queensland, Australia) using different pretreatment strategies," *RADIOCARBON,* Vol. 43. University of Arizona (2001).

Van Valkenburgh, Blaire, et al. "The Plio-Pleistocene Cheetah-Like Cat Miracinonyx inexpectatus of North America," *Journal of Verte-brate Paleontology* Vol. 10, No. 4 (Abingdon, UK: Taylor & Francis, for the Society of Vertebrate Paleontology, Dec. 20, 1990).

Varki, Ajit, and Nelson, David L. "Genomic Comparisons of Humans and Chimpanzees," *Annual Review of Anthropology* 36 (2007).

Villmoare, Brian William, et al. "Early *Homo* at 2.8 Ma from Ledi-Geraru, Afar, Ethiopia," *Science* Vol. 347, Is. 6228 (20 Mar 2015).

Vrba, Elizabeth. "Biostratigraphy and chronology, based particularly on Bovidae, of the southern hominid-associated assemblages: Makapansgat, Sterkfontein, Taung, Kromdraai, Swartkrans; also Elandsfontein (Saldanha), Broken Hill (now Kabwe) and Cave of Hearths," (Eds.) H. de Lumley and M.-A. de Lumley, *Pretirage, ler Cong. Internat. Paleo. Humaine,* Vol. II (1982).

Wallace, Alfred. *The World Of Life: A Manifestation Of Creative Power, Directive Mind and Ultimate Purpose*, New York: Moffat, Yard (1916).

Walsh, John E. *Unraveling Piltdown: The Science Fraud of the Century and Its Solution* (Random House, 1996).

Ward, C. V., et al. "Complete Fourth Metatarsal and Arches in the Foot of *Australopithecus afarensis,"* *Science* 331 (2011).

Ward, Carol, "Bipedal Foot Morphology," *Centre for Academic Research and Training/Anthropogeny* (https://carta.anthropogeny.org) (Retrieved March 21, 2021).

Werdelin, Lars. "Did Climate Change Shape Human Evolution?" Symposium, Columbia University's Lamont-Doherty Earth Observatory, Palisades, NY (April 19–20, 2012).

Werdelin, Lars and Lewis, Margaret E. "Temporal Change in Functional Richness and Evenness in the Eastern African Plio-Pleistocene Carnivoran Guild," *2PLoS One.* 8(3): e57944. Published online (2013).

White, Tim D., Asfaw, B. et al. "Pleistocene *Homo sapiens* from Middle Awash, Ethiopia," *Nature*, 423 (6491) (2003).

White, Tim, et al. "A New Kind of Ancestor: Ardipithecus Unveiled, Published by AAAS," *Newsfocus, Science Vol 326* (2 Oct 2009).

Whiten, Andrew and Erda, David. "The human socio-cognitive niche and its evolutionary origins," (Philosophical Transactions of the Royal Society Biological Sciences) vol. 367 (2012).

Whitfield, John. "Oldest member of human family found," *Nature— News Feature.* Springer Nature (2002).

Wilson, E. O. *The Creation: An Appeal to Save Life on Earth* (New York, W. W. Norton, 2006).

Wilson, E. O. *Diversity of Life.* W. W. Norton Company, (1992).

Wilson, E. O. *Consilience: The Unity of Knowledge.* Vintage Books (1999).

Wrangham, Richard. *Catching Fire: How Cooking Made Us Human* (New York, Basic Books 2010).

Wroe, Stephen, et al. "How to build a mammalian super-predator," *Zoology* (Jena) Vol. 111(3) (Mar 3, 2008).

Yeoman, Barry. "Billions to None—Why the Passenger Pigeon Went Extinct," *Audubon* (May–June 2014).

Zhu, Zhaoyu, et al. "Hominin occupation of the Chinese Loess Plateau since about 2.1 million years ago," *Nature* 559 (2018).

Online Resources

https://truenaturefoundation.org/aurochs. Accessed and verified May 2, 2021.

https://livingplanet.panda.org. Accessed and verified May 2, 2021.

https://www.youtube.com/watch?v=4NZ0JuDAyTo&t=3s. Accessed and verified May 2, 2021.

https://www.youtube.com/watch?v=GPogpX2z594. Accessed and verified May 2, 2021.

https://www.oldest.org/animals/tortoises. Accessed and verified May 2, 2021.

www.sciencedaily.com/releases/2018/10/181010141739.htm. Accessed and verified May 2, 2021.

www.savetheelephants.org/blog. Accessed and verified May 2, 2021.

https://youtu.be/MEmdNT1tEEw. Accessed and verified May 2, 2021.

https://youtu.be/p4xwRQSYlSk. Accessed and verified May 2, 2021.

https://youtu.be/bnT5vNE7LMI. Accessed and verified May 2, 2021.

https://www.nationalgeographic.org/media/tanzania-terror. Accessed and verified May 2, 2021.

https://www.youtube.com/watch?v=B6wKgENYLgA. Accessed and verified May 2, 2021.

https://carta.anthropogeny.org. Accessed and verified May 2, 2021.

https://humanorigins.si.edu/evidence/human-fossils/species/homo-rudolfensis. Accessed and verified May 2, 2021.

https://www.mothersinscience.com/trailblazers/jane-goodall. Accessed and verified May 2, 2021.

https://en.wikipedia.org/wiki/Not_even_wrong. Accessed and verified May 2, 2021.

Acknowledgements

THIS BOOK COULD NOT HAVE BEEN WRITTEN WITHOUT THE encouragement and assistance of the late Paul S. Martin.

My ex-wife and close friend Delphine du Toit also played an indispensable and essential role from its conception to its completion.

The book's publication was made possible by the encouragement and financial backing provided by my older son, Eric Edmeades. The technical and many other kinds of support I received from my younger son, Nik Edmiidz, were also of critical importance to this project.

Paul Hearnden made a big contribution to the book by donating his much-needed and much-appreciated editorial skills to its production.

Thanks, too, to Mikey Kershisnik for her expert and good-humored help in finalizing the editorial process. Any remaining glitches are my responsibility rather than hers, or that of members of her team.

I would like, finally, to express my gratitude to, and for, my partner Martha Brickman whose love and support made an enormously important contribution to the completion of this task.